# *Praise for* ***Education for Total Liberation***

"At the intersections of education and action, this important collection illuminates the importance of praxis to critical studies, offering readers a variety of entry points into the increasingly important realm of critical animal studies for liberation."

**Erik Juergensmeyer**, editor of *Green Theory and Praxis Journal*

"*Education for Total Liberation* is a must-read for activists, students and scholars. Finally, an entire book on critical animal pedagogy! This important text calls for a radicalising of what the field of critical animal pedagogy stands for and the different perspectives that exist across the field in both theory and practice. What an exciting book!"

**Lara Drew**, member of the Executive Board of Directors of the Institute for Critical Animal Studies

# Education for Total Liberation

RADICAL ANIMAL STUDIES
AND TOTAL LIBERATION

Anthony J. Nocella II
*Series Editor*

Vol. 2

The Radical Animal Studies and Total Liberation series
is part of the Peter Lang Education list.
Every volume is peer reviewed and meets
the highest quality standards for content and production.

PETER LANG
New York • Bern • Berlin
Brussels • Vienna • Oxford • Warsaw

# Education for Total Liberation

## Critical Animal Pedagogy and Teaching Against Speciesism

Edited by Anthony J. Nocella II, Carolyn Drew,
Amber E. George, Sinem Ketenci,
John Lupinacci, Ian Purdy, and J.L. Schatz

PETER LANG
New York • Bern • Berlin
Brussels • Vienna • Oxford • Warsaw

**Library of Congress Cataloging-in-Publication Data**

Names: Nocella II, Anthony J., editor.
Title: Education for total liberation: critical animal pedagogy
and teaching against speciesism / edited by Anthony J. Nocella II [and six others].
Description: New York: Peter Lang, 2019.
Series: Radical animal studies and total liberation; vol. 2
ISSN 2469-3065 (print) | ISSN 2469-3081 (online)
Includes bibliographical references and index.
Identifiers: LCCN 2018023768 | ISBN 978-1-4331-3435-7 (hardback: alk. paper)
ISBN 978-1-4331-3434-0 (paperback: alk. paper)
ISBN 978-1-4331-5788-2 (ebook pdf) | ISBN 978-1-4331-5789-9 (epub)
ISBN 978-1-4331-5790-5 (mobi)
Subjects: LCSH: Animal rights—Study and teaching. | Speciesism—Study and teaching.
Classification: LCC HV4708.E38 2018 | DDC 179/.3071
LC record available at https://lccn.loc.gov/2018023768
DOI 10.3726/b14204

Bibliographic information published by **Die Deutsche Nationalbibliothek**.
**Die Deutsche Nationalbibliothek** lists this publication in the "Deutsche
Nationalbibliografie"; detailed bibliographic data are available
on the Internet at http://dnb.d-nb.de/.


29 Broadway, 18th floor, New York, NY 10006
www.peterlang.com

This book is dedicated to all those exploited and murdered
in the name of capitalism and punishment, such as those
in prisons, cages, factories, tanks, and chained.

until all are free…

# *Table of Contents*

# Acknowledgments

The editors of this book would like to thank everyone involved with Peter Lang Publishing for believing in and supporting this project. We would like to thank everyone involved with the Institute for Critical Animal Studies, *Journal for Critical Animal Studies*, *Peace Studies Journal*, *Green Theory and Praxis Journal*, Save the Kids, Arissa Media Group, and Poetry Behind the Walls. We would also like to thank all of the contributors, Anne C. Bell, Constance L. Russell, Karin Gunnarsson Dinker, Helena Pedersen, Sean Parson, Meneka Repka, Richard J. White, Anita de Melo, and William E. Shanahan III. We would also like to thank nonhuman animals, Animal Liberation Front and Animal Liberation ACT, and radical educators everywhere. John would like to acknowledge the human and nonhuman members of his family, the students and faculty in the Cultural Studies and Social Thought program at Washington State University, the radical activists sacrificing life and freedom for liberation, and all the radical activist-educators fighting in classrooms and communities to abolish the logics of domination pervasive in schools. Anthony would like to thank Becky, Carey, Mark, Dawn, Keri, Kate, and Janine in the Department of Sociology at Fort Lewis College. He would also like to thank Mari, Chris, Michael, Terry, Jorg, Carlos, Cam, Cannon, Olivia, Logahn, Sierra, Brenna, Arissa, Erik, Chiara, Kara, Damon, Chad, Tracy, Don, Arash, Ahmad, Laurie, Torie, Daniel, Janice, John, Sidney, Kaidee, Emma, MeKayla, Mikayla, Cleo, his family, his professors at Syracuse University, Fresno Pacific University, University of St. Thomas and Jay, Ben, Brian, Jeremy, and many other friends he has organized with throughout the years for social justice!

# *Introduction: Examining the Nexus: Critical Animal Studies and Critical Pedagogy*

Carolyn Drew, Amber E. George, Sinem Ketenci, John Lupinacci, Anthony J. Nocella II, Ian Purdy, and J.L. Schatz

## *Introduction*

Critical animal studies (CAS) has increasingly grown as a scholarly movement over the past decade, after emerging out of critical thinkers who wished to dislodge the tenants of humanism. However, despite its growth, CAS has been conflated with other branches of critical theory and scholarship on speciesism. These other approaches include strictly utilitarian concerns surrounding human–animal relationships, embodied in thinkers like Peter Singer and pragmatic reformist organizations like the People for the Ethical Treatment of Animals (PETA). It also includes approaches that retreat purely into the ontological origins of epistemic constructions of the animal, embodied by Derridean deconstruction and academic scholarship. What distinguishes CAS from other approaches to human–animal studies is that CAS mandates theory be intimately tied to action with the intent of supporting total liberation from an intersectional perspective grounded in anarchism and supporting radical revolutionary direct action such as the Animal Liberation Front (ALF; Best & Nocella II, 2004). This is not to say any action advancing the cause of total liberation will be perfect or uncompromising when performed. At the same time, the endpoint of total liberation for all—both human and nonhuman alike—must not be sacrificed when weighing such concessions. The reasoning is that all too often small advancements for reform turn back the pace of more revolutionary change (Nocella II, White, & Cudworth, 2015). For CAS, advancing the liberation of nonhuman animals through using other human oppression as an absent referent, or at the sacrificing of individuals with disabilities, is unacceptable (Nocella II, Bentley, & Duncan, 2012). In short, while tactics may vary between CAS scholars, what remains consistent is working to dismantle the exclusionary circles of compassion and fighting against all forms of injustice. CAS is against academia and the academic industrial complex along with all forms of elitism, individualism, competition, capitalism, and exploitation, which academics promote and foster. One does not need to be an academic to be in the academic industrial complex/higher education, but rather a scholar. CAS and the Institute for Critical Animal Studies (ICAS) promote scholar-activists and activist-scholars, who are reflective and critical.

As one model for CAS, ICAS was formed in 2001 by Anthony Nocella II and Steve Best to bring activists and scholars together (ICAS, 2017). In 2007, Steve Best, Anthony J. Nocella, II, Richard Kahn, Carol Gigliotti, and Lisa Kremmerer shared from ICAS principles for a CAS that:

1. Pursues interdisciplinary collaborative writing and research in a rich and comprehensive manner that includes perspectives typically ignored by animal studies such as political economy.
2. Rejects pseudo-objective academic analysis by explicitly clarifying its normative values and political commitments, such that there are no positivist illusions whatsoever that theory is disinterested or writing and research is nonpolitical. To support experiential understanding and subjectivity.
3. Eschews narrow academic viewpoints and the debilitating theory-for-theory's sake position in order to link theory to practice, analysis to politics, and the academy to the community.
4. Advances a holistic understanding of the commonality of oppressions, such that speciesism, sexism, racism, ablism, statism, classism, militarism and other hierarchical ideologies and institutions are viewed as parts of a larger, interlocking, global system of domination.
5. Rejects apolitical, conservative, and liberal positions in order to advance an anti-capitalist, and, more generally, a radical anti-hierarchical politics. This orientation seeks to dismantle all structures of exploitation, domination, oppression, torture, killing, and power in favor of decentralizing and democratizing society at all levels and on a global basis.
6. Rejects reformist, single-issue, nation-based, legislative, strictly animal interest politics in favor of alliance politics and solidarity with other struggles against oppression and hierarchy.

7. Champions a politics of total liberation which grasps the need for, and the inseparability of, human, nonhuman animal, and Earth liberation and freedom for all in one comprehensive, though diverse, struggle; to quote Martin Luther King Jr.: "Injustice anywhere is a threat to justice everywhere."
8. Deconstructs and reconstructs the socially constructed binary oppositions between human and nonhuman animals, a move basic to mainstream animal studies, but also looks to illuminate related dichotomies between culture and nature, civilization and wilderness and other dominator hierarchies to emphasize the historical limits placed upon humanity, nonhuman animals, cultural/political norms, and the liberation of nature as part of a transformative project that seeks to transcend these limits towards greater freedom, peace, and ecological harmony.
9. Openly supports and examines controversial radical politics and strategies used in all kinds of social justice movements, such as those that involve economic sabotage from boycotts to direct action toward the goal of peace.
10. Seeks to create openings for constructive critical dialogue on issues relevant to Critical Animal Studies across a wide-range of academic groups; citizens and grassroots activists; the staffs of policy and social service organizations; and people in private, public, and non-profit sectors. Through—and only through—new paradigms of ecopedagogy, bridge-building with other social movements, and a solidarity-based alliance politics, is it possible to build the new forms of consciousness, knowledge, social institutions that are necessary to dissolve the hierarchical society that has enslaved this planet for the last ten thousand years. (pp. 1–2)

Since that time, ICAS has grown into an international organization that hosts annual conferences on several continents while keeping its intersectional focus and call for total liberation alive. While several schools of thought and activists have splintered away from ICAS to find their own niches, ICAS continues to serve as an outpost for new students to enter into and explore how best how to achieve the aim of CAS for themselves. In fact, despite its differences with other individuals and organizations, ICAS could not have come into existence without these perspectives. Their efforts allowed ICAS and its supporters to branch out and explore other avenues for resistance after fracturing apart. In many instances, while the differences are not insignificant, the coalitional benefit of varied strategies outweighs these smaller differences so long as the principle of a total liberation is upheld as the goal. As Richard Kahn and Brandy Humes (2009) explained:

> [T]otal liberation pedagogy, following the advances of multicultural educational theory, views oppression in systematic and complex terms, what Patricia Collins (2000) has termed the "matrix of domination." This not only allows for a more refined analysis of the ways in which power circulates throughout nature and culture, to the systematic advantage of some and disadvantage of others, but by increasing the number of epistemic standpoints from which to teach and learn we free a potential multitude of educational subjects from the culture of silence generated by the dominant mainstream pedagogical and political platforms. (p. 182)

Therefore, the principle of total liberation requires accepting competing subjectivities even when in disagreement in order to strategize against dominant frameworks and political orientations. The acceptance of seemingly mutually exclusive possibilities can mobilize one to challenge current power structures as long as these differences in approach are individualized and work to disallow such differences to get in the way of an overall strategy of total liberation.

Since falling short is always inevitable, the aim of CAS is not to let the perfect be the enemy of the good or strictly to define what counts as CAS scholarship. Nevertheless, it is important to distinguish CAS from strategies that are fundamentally at odds with total liberation. To this end, CAS is defined by its courage to do more and not succumb to the good as the excuse for not pushing for the perfect. This requires a willingness to reflect on the interlocking matrixes of domination that persist even as we struggle to resist. Despite any setbacks we may encounter, we must be willing to try again tomorrow until we have achieved a world where all are free no matter how long it takes. If along the way, total liberation is abandoned as the primary objective, the struggle becomes disempowered since being good turns into an excuse for being good enough and not doing more. Furthermore, this causes scholarship to lose its critical edge and activists to be co-opted into the nonprofit industrial complex. Put simply, CAS as a field teaches total liberation through the praxis of turning scholarship into activism and turning the actions of organic intellectuals into academic knowledge. To this end, "total liberation pedagogy does not seek to just destabilize human power in the abstract, but roots this in the need to support cultural and political practices that actively seek to overthrow speciesist relations across society" (Kahn & Humes, 2009, p. 182). Hence, scholarship cannot reside in the academy alone. Instead, scholarship must empower those who learn from it to integrate that knowledge into their day-to-day lives as meaningfully as possible. The aim of this book is to do precisely that.

Sadly, much of the work on critical praxis and social justice that aims for liberation fails to consider non/humans as a subjectivity worth considering. This is unfortunate because, when organizing, it fuels "more profits for the corporations that own or contract with factory farms…[used at] the food services at the hotels at which every social justice conference I've ever attended were hosted" (Gorski, 2014, p. 8). Again, to be clear, we recognize not everyone can avoid factory farms and industrial agriculture due to various socioeconomic reasons. However, when organizers promote events at places where vegetarian and vegan options are available at comparable or lower costs, to still serve flesh is giving up on total liberation. Furthermore, the sacrifice of literal lives during an academic event that supposedly calls for social justice is not a minor area of disagreement. Gorski (2014) explained:

> There's a price to pay for sitting atop an abusive industry, and part of that price is representing that industry's atrocities[, which include being]…the biggest source of greenhouse gas emissions in the world (Goodland & Anhang, 2009)—bigger, in fact, than all other sources combined. But that's not all, because the havoc factory farms wreak is varied and extensive, and it targets some of the most marginalized beings in the world. … Farm animals…[are] slaughtered at a rate of about 1 million per hour…[while serving as] a form of violence against humans. (p. 7)

To this end, CAS is an important starting point for organizing around social justice since it includes speciesism within its intersectional understanding instead of leaving it as an afterthought. This is important now more than ever because

> we live in a time of a mass species extinction event such as we have not witnessed on the planet for nearly 65 million years. The zoöcidal eradication of unprecedented numbers of mammals, amphibians, reptiles, birds, fish, insects, and other animals that is now fully underway…[driven by] the exponential growth…of the industrial factory farm model of animal agriculture. (Kahn & Humes, 2009, p. 182)

If one were to ignore these facts, then any progress made by activists would be locked into a spiral of inevitable failure that only strengthens the corporate and governmental entities that solidify oppression in the first place. The goal of CAS in general and ICAS specifically is to ensure that doesn't happen.

In 2010, in the book *Animals in Schools: Processes and Strategies in Human-Animal Education,* Helena Pedersen brought CAS explicitly into the field of education. Richard Kahn's (2010) *Critical Pedagogy, Ecoliteracy, & Planetary Crisis: The Ecopedagogy Movement* brought the principles of critical pedagogy into consideration and the strong argument for pedagogical consideration in schools. Around the same time, ICAS began to explicitly foster the importance of developing pedagogical examples and experiments of CAS work in schools and education more broadly. The *Green Theory and Praxis Journal* and *Journal for Critical Animal Studies* have been devoted over the years to CAS scholar-activist, educators sharing how they teach within a CAS framework. More recently, stemming from work in education research and bringing posthumanism together with the emerging CAS, Helena Pedersen (2011) introduces what she calls a "critical posthumanist approach" (p. 67). Following that work, Gunnarsson Dinker and Pedersen (2016)—in a chapter in a handbook of alternative education—introduce what they coin as "critical animal pedagogies." They pose for educators and education researchers the challenge of seeking alternatives to dominant human-centric models of education, and they propose that a critical animal pedagogy illuminates, often overlooked and undervalued, knowledges of human–animal relations. In other words, Pederson and Gunnarsson Dinker invite scholar-activist educators to examine what education can be if humans are not the center of focus and understood as superior.

We hope that the collection of chapters in this volume provides readers with insight into how CAS scholar-activists are grappling with the praxis of critical animal pedagogies rich with a variety of tactics and styles. However, the chapters will also be useful for those who are not immediately interested in CAS as a field. Due to the intersectional orientation of CAS, its pedagogy necessitates that all the authors included in this collection think beyond just the non/human animal. Drawing from Kho (2016), this is done beyond a mere paralleling of oppression but by

> going to the root of these oppressions conceptually. … What makes the physical violation of these bodies possible is their citizenship to the space of "the other" or the "sub-human." … Comparing and contrasting the literal/physical violations these subjects experience misses the conceptual boat because the reason why they are each oppressed is precisely because they ALL are citizens of the same sub-human space. … They are not being oppressed because they are "like" each other. They are being oppressed because they have been labeled as "less-than-human" where human is defined as the superior and ideal white species. (para. 13)

The conceptual maps presented within this book serve as a method for attacking the oppression that solidifies itself through a form of liberal humanism. The varied tactics deployed in these chapters serve as testament to a plurality of ways to embody the overall strategy of total liberation and tangible praxis.

## *Overview of the Book*

We have organized this book to help the reader ease into understanding not only CAS scholarship but also the pedagogy that informs its practice. The first chapter, "Unmasking the Animal Liberation Front Using Critical Pedagogy: Seeing the ALF for Who They Really Are," is a selection by Anthony J. Nocella II from the book *Terrorists or Freedom Fighters? Reflections on the Liberation of Animals* (2004) coedited by Steve Best and Anthony J. Nocella II. This chapter dispels myths about the ALF and articulates the strong allyship between CAS scholar-activism and radical direct action organizations like the ALF. Furthermore, Nocella II highlights connections between the ALF, critical pedagogy, and direct action rooted in love and liberation for all species.

Chapter Two, "Beyond Human, Beyond Words: Anthropocentrism, Critical Pedagogy, and the Poststructuralist Turn," is reprinted with permissions from the *Canadian Journal of Education* where the authors, Anne C. Bell and Constance L. Russell (2000), articulated with precision the tendencies of critical pedagogical efforts in reproducing anthropocentrism. In this chapter Bell and Russell propose critical possibilities for critical environmental education in addressing the limitations often found in poststructural approaches to critical pedagogy.

In Chapter Three, "Critical Animal Pedagogy: Explorations toward Reflective Practice," the authors, Karin Gunnarsson Dinker and Helena Pedersen, make the astute argument that education actively (re)produces injustices and inequities in society. Drawing from Pierce's (2013) work on biocapitalism and focusing on the commodification of life, Gunnarsson Dinker and Pederson expose what MacCormack (2013) refers to as a *war* between humans and other species. Gunnarsson Dinker and Pederson then propose a possibility for critical animal pedagogy for disrupting the normalcy of education in favor of students and animals jointly creating new educational possibilities.

In Chapter Four, "Our Heroes Need to Wear Ski-Masks: The Animal Man, and the Animal Liberationist Hero in Comics," Sean Parson argues that comics, something the reader might usually identify with conservative narratives which uphold societal norms, can serve as forms of educational spaces for discussions around animal liberation generally. Parson discusses how although the media is often only understood by the left as spaces of propaganda, these same spaces can be places of education outside of the classroom. He claims that despite its use as a tool of social control, popular culture, through media, can serve as a "critical response" to the power structures that dominate and control the status quo. In dipping into the history of the comic, in particular the super-hero genre, Parson shows us how the narratives though often seemingly supporting the status quo actually open up a space for critical response in the tension between the normative values on the one hand and the often difference or otherness of the main character/s on the other hand. Parson explores the series *Animal Man* paralleling the development of its narrative, and its turning points, with that of discussions around CAS and total liberation. Through an exploration of *Animal Man* Parson also explores the interconnectedness of oppression and in doing so enables the reader to reflect on the broader movement of total liberation through the narrative of comics such as *Animal Man* and animal liberation.

In Chapter Five, "Teaching to End Human Supremacy↔Learning to Recognize Equity in All Species," John Lupinacci offers an overview of an ecocritical framework and how human supremacy is learned. The chapter explores the possibilities of what Lupinacci refers to as (un)learning anthropocentrism, both inside and outside of schools. Drawing from CAS, anarchism, and ecojustice education, Lupinacci draws attention to the potential for educators to interrupt anthropocentrism and possible alternatives for educators to teach toward social justice and sustainability. Illustrating how dominant cultural assumptions of human supremacy contribute to social suffering and environmental degradation, in this chapter Lupinacci emphasizes the pedagogical potential for critical animal pedagogies aimed at ending human supremacy. The author concludes by providing guiding steps that can be utilized in order to teach in support of social justice and sustainability.

In Chapter Six, "Intersecting Oppressions: The Animal Industrial Complex and the Educational Industrial Complex," Meneka Repka brings CAS into K–12 classrooms. Repka, explaining that the animal industrial complex and the educational industrial complex are too often examined separately, addresses the similarities between these complexes and then focuses in on the nuanced ways in which they are intertwined. In turn, she illustrates how the animal agriculture industry and systems of public schooling work synergistically to maintain their positions of power. Ultimately, Thirukkunaran illustrates how it is that schools become concentrated sites of intersecting human oppressions and offers optimism and hope that educators can, and ought to, take direct action through their teaching toward dismantling schools in favor of critical animal pedagogies—living curriculums in support of total liberation.

In Chapter Seven, "'We are one lesson': Some Reflections on Teaching Critical Animal Geographies in the University to Undergraduate Students," Richard White shares how critical animal pedagogies offer the potential for questioning dominant human–nonhuman animal relations and encouraging a new intersectional consciousness to emerge. White describes this consciousness as such that recognizes and challenges oppressive, exploitative, and violent relationships between humans and other animals. The chapter explores the reflective accounts from how White has used anarchist praxis, and CAS, to inform critical animal geographies in the classroom.

In Chapter Eight, "Teaching the Animal in Foreign Languages: An Ecopedagogical Approach," Anita de Melo shares insight from her experience of teaching foreign languages. As an instructor of Portuguese in the United States, de Melo asserts that the learning of a foreign language can also be an act of developing sociopolitical awareness. With this at the core of her teaching, de Melo explains how she helps students develop consciousness for the plight of nonhuman animals and helps students understand the importance of interdependence and an awareness of the planetary condition. She shares an example from teaching a foreign language aimed at educating students on how both Portuguese and English choice of words for insults and idiomatic expressions rely heavily on derogatory and speciesist terms toward nonhuman animals—and connects these examples to the pervasiveness of the exploitation of nonhuman animals and the environment.

In Chapter Nine, "Activist Education for Animal Ethics: The Imperative of Intervention in Education on the Non/Human," J.L. Schatz argues that total liberation requires more than "theoretical discussion in the classroom." He argues that speciesism can only be challenged in the classroom by taking it further than the traditional papers and grading typical in most higher education institutions. Schatz gives the reader a glimpse into the journey he and the students took as they wrestled with the ethics of eating a non/human animal. Schatz points out that although few in his classroom denied the intrinsic value of the non/human animal, little had changed in regard to the percentage of students becoming vegetarian or vegan. To try and address this contradiction Schatz introduced the veg pledge where at the conclusion of his unit on non/human liberation the students were offered extra credit points for committing to going vegetarian or vegan. The chapter reflects on the first steps of transformation of those who took the pledge in merging theory (their classroom discussions and papers) with practice (changing to a vegetarian or vegan diet). Schatz urges readers to make teaching spaces about more than merely regurgitating information without internalizing it.

In Chapter Ten, "Building Alliances for Nonhuman Animals Using Critical Social Justice Dialogue," Amber E. George explores how activists and scholars can navigate the conflicts that arise from differing perspectives about animal liberation within CAS. Specifically, she explores how dialogue can help humans work through the complicated nature of identity politics that often causes misunderstanding based on one's racial identity, disability status, religion, species, gender, and so on. She argues that learning how to adequately address such conflict can improve nonhuman liberation and also other social justice efforts.

In Chapter Eleven, William E. Shanahan III closes the anthology of essays in this book with "Decolonizing Critical Classrooms, Decentering Animal Pedagogies." In this chapter Shanahan III offers a departure from the traditional writing style followed by a majority of scholar-activists. Shanahan III, recognizing how critical pedagogy upends foundational assumptions about classroom power, structures, discourse, relationships, challenges these generational hierarchies and demands a radical reconfiguration of teaching and learning—most notably by fundamentally decentering the roles of teacher and canonical knowledge. In doing so, Shanahan III criticizes how at times even CAS has fallen short in its attempts to define itself as a discipline. However, by altering the methodological terrain it can be possible to move beyond the very logocentrism of the academy that has perpetuated humanist violence.

Ultimately, our hope is that upon completing these chapters you as a reader will not only have a better understanding of what CAS is, but, perhaps more importantly, have internalized different methods of approaching education in order to advance liberation. Certainly, not every teaching strategy and activist tactic in this book will be good for everyone in every class or activist organizing session. However, many of the strategies the authors of this book have deployed in their organizing and teaching are readily portable to other subject matters and situations. At the same time we believe that the next wave of CAS and liberation must come from you. By taking these chapters as a springboard for how to develop further ideas on intersectional organizing, or more practically engaged education, we can continually push forward to overcome past mistakes and ensure no human or animal gets left behind for some short-term gain. Together we can break down the disciplinary boundaries of the separate areas of scholarship, styles of writing, and methods of organizing in order to alter the bedrock that grounds human supremacy. Education is a necessary tool in this struggle. Action is too. Through a critical praxis that combines both that bedrock can be shattered and a world of total liberation can be built. The more we collectively address anthropocentrism and the failures of many social justice advocates to recognize the complex ways oppression depends on the exploitation of non/human animals, the more we recognize that there is no one way of taking action—as illustrated by the diversity of actions taken in the chapters of this book. Along that same line of thinking and from the shadows of solidarity, it is paramount that we work together as allies to confront the systemic roots of oppression on all our respective fronts.

In other words, we have a responsibility, as scholar-activist educators committed to CAS, to support the oppressed in whatever capacity we can. Together our goal is to break the will of the oppressor and always seek total liberation.

## References

Bell, A. C., & Russell, C. L. (2000). Beyond human, beyond words: Anthropocentrism, critical pedagogy, and the poststructuralist turn. *Canadian Journal of Education/ Revue canadienne de l'education*, *25*(3), 188–203.

Best, S., & Nocella, A. J., II. (2004). *Terrorists or freedom fighters? Reflections on the liberation of animals.* New York, NY: Lantern Books.

Best, S., Nocella, A. J., II, Kahn, R., Gigliotti, C., & Kremmerer, L. (2007). *Introducing critical animal studies.* ICAS. Retrieved from http://www.criticalanimalstudies.org/wp-content/uploads/2009/09/Introducing-Critical-Animal-Studies-2007.pdf

Goodland, R., & Anhang, J. (2009). Livestock and climate change: What if the key actors in climate change are cows, pigs, and chickens? World Watch, November/December issue, 10–19.

Gorski, P. C. (2014). Consumerism as racial and economic injustice: The macroaggressions that make me, and maybe you, a hypocrite. *Understanding and Dismantling Privilege*, *4*(1), 1–21.

Gunnarsson Dinker, K., & Pedersen, H. (2016). Critical animal pedagogies: Re-learning our relations with animal others. In H. E. Lees & N. Noddings (Eds.), *The Palgrave international handbook of alternative education* (pp. 415–430). London: Palgrave Macmillan UK.

Institute for Critical Animal Studies (ICAS). (2017). "About." Retrieved from http://www.criticalanimalstudies.org/about/

Kahn, R. (2010). *Critical pedagogy, ecoliteracy, & planetary crisis: The ecopedagogy movement.* New York, NY: Peter Lang.

Kahn, R., & Humes, B. (2009). Marching out from Ultima Thule: Critical counterstories of emancipatory educators working at the intersection of human rights, animal rights, and planetary sustainability. *Canadian Journal of Environmental Education*, *14*(1), 179–195.

Kho, A. (2016, February 24). Why animal liberation requires an epistemological revolution. *Aphro-ism.* Retrieved from https://aphro-ism.com/2016/02/24/why-animal-liberation-requires-an-epistemological-revolution/

MacCormack, P. (2013). Gracious pedagogy. *Journal of Curriculum and Pedagogy*, *10*(1), 13–17.

Nocella, A. J., II, Bentley, J. K. C., & Duncan, J. (2012). *Earth, animal, and disability liberation: The rise of the eco-ability movement.* New York, NY: Peter Lang Publishing.

Nocella, A. J., II, White, R., & Cudworth, E. (2015). *Anarchism and animal liberation: Essays on complementary elements of total liberation.* Jefferson, NC: McFarland.

Pedersen, H. (2010). *Animals in schools: Processes and strategies in human-animal education.* West Lafayette, IN: Purdue University Press.

Pedersen, H. (2011). Release the moths: Critical animal studies and the posthumanist impulse. *Culture, Theory and Critique, 52*(1), 65–81.

Pierce, C. (2013). *Education in the age of biocapitalism: Optimizing educational life for a flat world.* New York, NY: Palgrave Macmillan.

# *Part I: Foundations*

# 1. *Unmasking the Animal Liberation Front Using Critical Pedagogy: Seeing the ALF for Who They Really Are**

ANTHONY J. NOCELLA II

The current global political climate is steeped in fear and rhetoric about terrorism and security. The 21st century began with drastic shifts in U.S. policies in the name of national security, which has been used as a cover for the repression of nonviolent dissent and the violation of civil liberties. We have entered a neo-McCarthyist period rooted in witch hunts against activists and critics of the ruling elites. The terms and players have changed, but the situation is much the same as in the 1950s: the terrorist threat usurps the communist threat, Attorney General John Ashcroft dons the garb of Senator Joseph McCarthy, and the Congressional Meetings on Eco-Terrorism stand in for the House Un-American Activities Committee (Best & Nocella, 2004). Now as then, the government informs the public that the nation is in a permanent state of danger, such that security, not freedom, must become our overriding concern. As before, the state conjures up dangerous enemies everywhere, not only outside our country but, more menacingly, ensconced within our borders, lurking in radical cells. The alleged dangers posed by foreign terrorists are used to justify the attack on "domestic terrorists" within, and in a hysterical climate the domestic terrorist is any and every citizen expressing dissent.

These changes have occurred as recent as U.S. President Bush illegally signing off on allowing the National Security Agency (NSA) to spy on citizens within the U.S., in the name of counterterrorism.[1] Along with the FBI

* Parts of this article have been directly taken from the Introduction to *Terrorists or Freedom Fighters? Reflections on the Liberation of Animals* (Lantern Books, 2004) coedited by Anthony J. Nocella, II and Dr. Steve Best. These sections have been reprinted with the publisher's permission.

roundup of six individuals on December 7, 2005, and on January 13, 2006, the arrest of three individuals in Auburn, California, the FBI and federal government have ratcheted up the war on activism.[2] On February 6, 2006, the SHAC 7 trial began in Trenton, New Jersey (Best & Kahn, 2004). This historical case centers on activists (animal liberation activists) being accused under the Animal Enterprise Protection Action (AEPA) of 1992 of protesting across U.S. state borders via the internet. The AEPA protects corporations that conduct business which tests on, kills for consumption (food or clothing), or uses for entertainment purposes nonhuman animals from public dissent.[3] From examples such as these, it is clear that there is a national witch hunt currently taking place. Where in the 1960s and 1970s, the FBI hunted down radical social groups such as the Black Panther Party, American Indian Movement, and antiwar activists (Abu-Jamal, 2000; Churchill, 2003; Churchill & Vander Wall, 2002a, 2002b; Jones, 1988, 1998; Peltier, 1999; Shakur, 1987), in the last decade or so they have begun targeting the animal liberation and environmental movements, particularly anyone supportive or suspected of involvement in the Animal Liberation Front (ALF) (Best & Nocella, 2004, 2006), which the state has termed the top "domestic terrorist" threat.[4] The ALF is an effective, decentralized, autonomous organization that in its actions provides a clear and compelling critique of corporate capitalist society. The ALF is any individual or group in any area of the world who at any time decides to strike against animal exploitation in the name of animal rights while following ALF guidelines.[5] Given the decentralized and anonymous nature of ALF actions, the ALF in principle is not about authority, ego, heroism, machismo, or martyrdom; rather, it is about overcoming hierarchy, patriarchy, passivity, and politics as usual so that creative individuals can dedicate themselves unselfishly to the cause of animal liberation. The structure and philosophy of the ALF thereby has some key affinities with anarchism and radical feminism.[6]

It cannot be denied that ALF actions have caused millions of dollars of damage in economic sabotage, and thus, the group represents a threat. A threat is correct, but to what is the important question. A threat to business like usual, where People of Color and the poor, nonhuman animals, and the nondomesticated world are exploited for the mighty dollar. The ALF represents no direct or overt threat to the U.S. government or to people, like the many right-wing groups that have virtually disappeared from the Department of Homeland Security (DHS) terror lists, despite targeting and threatening human lives. Rather, ALF actions threaten corporate abusers of the land, water, air, and animals. Every time they act, the lies and inequities contained within our current system of economic governance are laid bare.

If one is looking for groups with which to compare the ALF, the proper choice is not Al Qaeda or Saddam Hussein's Republican Guard, but rather the Jewish anti-Nazi resistance movement and the Underground Railroad. The men and women of the ALF pattern themselves after the freedom fighters in Nazi Germany who liberated war prisoners and Holocaust victims and destroyed equipment—such as weapons, railways, and gas ovens—that the Nazis used to torture and kill their victims. Similarly, by providing veterinary care and homes for many of the animals they liberate, the ALF models itself after the U.S. Underground Railroad movement, which helped fugitive slaves reach free states and Canada. Whereas corporate society, the state, and mass media brand the ALF as terrorists, the ALF has important similarities with some of the great freedom fighters of the past two centuries and is akin to contemporary peace and justice movements in its quest to end bloodshed and violence toward life and to win justice for other species.

## A False Perception

For too long opponents of illegal political property destruction tended to uncritically define it as violence and reject it as inherently wrong on this ground.[7] Their argument assumes the form of a classic syllogism: (1) property destruction is violence, (2) violence is always wrong, (3) therefore, property destruction is wrong. These critics rarely define what they mean by "violence," they dogmatically cling to the pacifist positions of Gandhi and King, and they make unqualified universal judgments that violence is always wrong and never works politically to achieve the goal goals of liberation.

Advocates of the principled critique believe that illegal actions and "violence" are unnecessary for a cause strong enough to prevail on the logical arguments supporting it. Peter Singer, for example, affirms "animal liberation" as a just cause, so long as it remains "nonviolent."[8] Violence can only beget more violence, he argues, recommending that animal liberationists emulate Gandhi and King in their goal to divest themselves of hatred, anger, and the will to revenge. Singer thinks that direct action is most effective when it brings results other tactics cannot and uncovers evidence of extreme animal abuse that awakens public understanding about the plight of animals. As an example of a just and effective raid, he points to the ALF break-in at the University of Pennsylvania head injury research laboratory, which exposed a truth never meant to be seen by the public. Singer argues that to stop or reduce animal suffering "we must change the minds of reasonable people in our society. ...The strength of the case for Animal Liberation is its ethical commitment; we occupy the high moral ground and to abandon it is to play into the hands of those who oppose us. ...The wrongs we inflict on other species are...[undeniable] once they are seen plainly; and it is in the rightness of our cause, and not the fear of our bombs, that our prospects of victory lie."[9] The motto here is not Burn Baby Burn, but Learn Baby Learn.

## *In Defense of the ALF*

While Singer and many others appeal to the "minds of reasonable people," the ALF believes that far too many are unreasonable and closed-minded, rendering the force of reason and persuasion insufficient. Industries and the state have strong institutional and monetary biases against justice for animals that no amount of persuasion or education is likely to change. Those who champion education and legislation as the sole tools of struggle project a rationalist belief that discounts the irrational forces often ruling the human psyche, the sadistic pleasure all too many derive from torture and killing, the deep psychological mechanisms human beings use to resist change and unpleasant realities, the mechanisms of detachment and compartmentalization that allow them to ignore the enormity of animal suffering, the vested interests they have in exploiting animals, and their identities as members of a species they believe is the preordained master of the Earth.

Unavoidably, animal liberationists are caught in a war of publicity and propaganda, and they must defeat the mendacity of the state and animal exploitation industries to fight for the hearts and minds of the people. While public opinion may indeed be secondary to the impact direct action can have on an industry, and potential negative media coverage should not deter activists from sabotage operations, it is a tactical mistake to act as if public thinking were irrelevant.[10] If negative images of ALF actions prevail, industries will win support, liberationists will lose sympathy, and few will protest when the state pounces on the ALF with fierce repression.

Unfortunately, many people believe that the ALF is a violent organization that grew out of anger and hate. In order to understand the ALF, one must sidestep this and other stereotypes, generalizations, and preconceived mindsets that for the most part have been developed and reenforced by mass media, capitalist ideologues, and legal authorities. Through a more open and informed approach, one will find that the ALF grew out of love for all life, a perspective strongly supported by the "ALF Guidelines," which state that it is crucial "...to take all necessary precautions against harming any animal, human and non-human."[11] The misconception that the ALF is a "violent" group can be avoided by understanding the group's origins and the reasons and motivations for their actions. I suggest a process the public can use to better understand the ALF and their actions.

## *Learning to Understand the ALF*

To truly understand those labeled as "terrorists," we must go beyond this fear and rhetoric of the "War on Terrorism" (which is nothing more than a war on ideas which resist oppression in all forms and global capitalism), packaged, promoted, and propagated by the U.S. Bush administration and fellow juggernauts like Tony Blair of England and seek a true understanding of these groups labeled as terrorists, in this case the ALF. We must seek to understand their goals and agendas, and to find out how to relate and engage in their struggle for liberation, which the powers that be fear and wish to destroy. The complex issue of terrorism in the 21st century calls for complementary approaches of analysis and transformation, therefore we must seek an approach that can not only unmask the ALF, but dethrone and stand down the multibillion dollar propaganda machine of the "War on Terrorism." This then stresses Jacques Derrida's point that language is a field of social control and to deconstruct it one must understand the relation it has to each other for nothing exists separate and on its own. He stresses the complexity and interconnectedness of this so-called "War on Terrorism" as being not a binary (of good vs. evil), but a "multinary," which must be examined from a diversity of vantage points, for each position or lens will show and discover something new and important (Derrida, 2001).

What better approach to fight and unveil the complex and interwoven lies of this machine than one that fights for justice and the oppressed, is critical, and promotes education as a form of transformation and not violent. *Critical pedagogy* is just that approach, which is based on praxis, process, and being a process for liberation (Freire, 1997; Kanpol, 1999; Kincheloe, 2004; Lather, 2001; McLaren, 2006). Critical pedagogy owes much gratitude to the founder of critical pedagogy, Paulo Freire, an internationally renowned educator, who bridged the gap between the revolutionaries and education. In examining Freire's seminal book, *Pedagogy of the Oppressed* (1997), one can understand the importance of striving for an engaging educational experience in the classroom or even learning about the revolutionary/liberation groups, societies, and collectives. Critical pedagogy strives for the teacher to step out of their authoritarian position, while holding onto authority as a facilitator and hit the streets with their students to analyze and engage in reality of life's many experiences. This approach based on social justice and fighting for the oppressed demands one to be critical especially of history, therefore examining the mass media, common historical texts, and advertising (Freire, 1985; Giroux, 1988, 1997; Herman & Chomsky, 1988; Irwin, 1996; Kanpol, 1999; Kincheloe, 2004; Loewen, 1995; Parenti, 1993; Shor, 1992; Zinn, 1999; Zinn & Arnove, 2004). Furthermore, critical pedagogues stress the active engagement with social justice as an activist, rather than a detached reader of history, to understand history we much engage in it. A novel might be capturing and poetic, but no one can tell another exactly why and how one feels about an experience without living it. Therefore, it is not merely about researching social problems, but engaging with them and fighting for social justice and stepping out of the safety of the classroom (or ivory tower) with one's students and into the streets (Gramsci, 1989).

The approach one adopts for understanding the ALF is the first and most crucial step in opening dialogue. To understand ALF, it is best to engage in a *critical pedagogy* approach, as understanding cannot always be achieved through a rigid barrier of "objectivity." "Critical pedagogy is that form of education which emerges from critical compassion; a transcendence of the emotional and the intellectual; the heart and mind learn to see and know in new ways."[12] In examining Paulo Freire's book, *Pedagogy of the Oppressed* (1997), one will find that educating is not just a technique in which one can achieve certain results. Critical pedagogy involves the dissolving of formal boundaries and requires a relationship where questions arise and sharing is included in the dialogue. The formal roles of the "teacher" and "student" relationship are absent in the critical pedagogy approach and replaced by the "syllabus of experience." It should be noted that, Richard Kahn[13] author of numerous articles and scholar of *ecopedagogy* (founded by Paulo Freire) can argue that connecting critical pedagogy to animal liberation is a form of ecopedagogy (Kahn, 2003).

The ALF are unique in that they are decentralized and made up of disparate cells with their own organizational structure, ideologies, and sociopolitical positions, which only have one thing in common: their guidelines and goal, to liberate animals. In understanding the ALF one must know that they do not act solely on emotions, but engage in critical thought to find the best step to achieve liberation, and are motivated by an acute sense of justice that, in principle at least, informs Western "democracies."

Furthermore, to understand the ALF is to grasp their history, culture, and mission. When one understands not only his/her own position, but also the position of the ALF member, one will truly be able to understand the motivations and reasoning of why the ALF exists. This approach does not suggest that the actions of the ALF are right, but rather that if the person uses critical pedagogy he/she is more likely to understand the ALF members' perspective.

In addition, it is not enough to physically experience the "subject"; it is also important to have one's mind and spirit involved in the experience. bell hooks notes that Vietnamese Buddhist monk Thich Nhat Hanh approaches this kind of pedagogy by emphasizing the union of mind, body, and spirit (hooks, 1994). An infiltrator of the ALF might be in a cell for a number of years and feel that he or she understands the ALF, but the essence of critical pedagogy is what Hanh explains as the necessity of the union of the mind, body, and soul, which all have to be engaged in the experience of understanding. One can begin to understand and appreciate ALF's actions through an intellectual understanding of justice, an emotional understanding of animals' suffering, or a spiritual understanding of the unity of all life. When all of these powers are engaged, critical pedagogy becomes possible.

Paulo Freire seems to suggest that critical pedagogy leads to the educator's or in this case the researcher's own liberation (Freire, 1997; Freire & Macedo, 1998; McLaren, 2000, 2006) through an experiential form of education that strives for enlightenment or what Gandhi might refer to as the state of seeking the truth. Gandhi believes that the search for truth is the ideal purpose of life. He explains that the struggle to free India was in fact the search for truth on a mass level—and suggests that it is only possible to be in a state of truth if one is willing to give up one's own freedom to stand up to an unjust act or law, or to save or free another being. Such people as St. Francis of Assisi, Henry David Thoreau, and Martin Luther King Jr. would also favor this perspective. Dr. King stated it best: "I became convinced that non-cooperation with evil is as much a moral obligation as is cooperation with good" (King, 2001). When the oppressed animals—to whom the ALF member is connected in a holistic and emotional manner—are freed, then part of the body and spirit of the ALF member is freed as well. For example, after a liberation by an ALF member or cell, not only do the animals experience freedom, but the ALF member experiences emancipation—*from* fear, alienation, and perhaps guilt (at not doing anything or enough), and *for* spiritual development. That is why it is common for a member of the ALF to cry with happiness after liberating an animal. The liberated animals and the ALF member all achieve a piece of the truth of wholeness and union.

The essence of performing an act in the name of the ALF is that love must be present in one's heart. This love allows one to act with respect for all life (even those who exploit animals) and to use one's intellect to its fullest potential.

- Have acts been committed in the name of the ALF by people in a negative or hostile state? Yes.
- Should activists in the wrong frame of mind be questioned by animal liberation and rights advocates? Yes.

How would one prevent such acts? One should never let go of the true meaning of the ALF, which is to respect all life (the fundamental principle of the animal rights movement as a whole), to the extent that one feels oppressed due to another's oppression. While many say that activists should use their anger and hate toward the enemy, I say it is better to emulate individuals like Jesus, Gandhi, Cesar Chavez, or other great peacemakers, and redirect anger and hatred into a state of love.

But one should never confuse nonviolence with weakness. King and Gandhi, though promoters of nonviolence did not favor cowardice or walking away from conflict (Gandhi, 1993; King, 2001). The ALF acts with strength and daring, yet stops at violence, because it acts from love. Do not be confused by the media or law enforcement, which is driven by U.S. Congress and corporations on how to think and act; this group promotes nonviolence and freedom for all. Their actions are revolutionary and sometimes "extreme" in resorting to property destruction and breaking the law, but to the right-thinking person these acts should pale in comparison to the truly extreme actions that involve not damaging things but injuring and taking life and profiting from killing and death—the routine actions of animal exploitation industries such as vivisection laboratories, zoos, circuses, factory farms, and slaughterhouses. It is only when all people understand that love will create love, and hate will only create hate, that all will be liberated. As the Quakers, Martin Luther King, Jr., Mohandas Gandhi, and the ALF believe, love will light the path to liberation (Gandhi, 1993; King, 2001; West, 1992).

## *Motivated by Love and Acting Nonviolently*

Resolved not to harm living beings, motivated by love, empathy, compassion, and justice, animal liberationists are the antithesis of the "terrorists" that government, industries, and mass media ideologues impugn them to be. They are not violent aggressors against life; they are defenders of freedom and justice for any enslaved species. They uphold rights not covered by law, knowing that the legal structure is defined by and for human supremacists. The goal of the ALF is not simply to liberate individual animals here and there; it is to free all animals from every form of slavery that binds them to human oppressors. The ALF, like the animal rights movement as a whole, is attacking the entire institutional framework of animal exploitation along with the domineering values, mindset, identities, and worldviews of the human species.

Thus, if one wants to understand the ALF, one must transcend the false rhetoric of "terrorism" and approach the real purpose of its struggle—animal liberation—through the method of critical pedagogy (Freire, 1985). One can thereby strive to understand the liberators' motivations as rooted in a concern for the suffering of nonhuman animals and for peace. For when you engage in understanding the ALF, you will understand the need for the liberation of all life.

## Notes

1. http://www.cnn.com/2005/POLITICS/12/17/bush.nsa/, February 15, 2006.
2. All of them were arrested on charges in relation to economic sabotage from the historic action at Vail, Colorado, where a ski lodge was burnt down causing 12 million dollars worth of damage, to burning down logging mills outside of Eugene, Oregon. The FBI, ATF, and local law enforcement arrested them as if they all were involved together, but it is a fact that most of the individuals arrested do not know each other and have not been in certain areas when these actions occurred. The facts are slim and shady at best, but they need someone to take the blame for these actions, and they are striving to break the back of radical resistance movements in the U.S., but so far they have been unsuccessful. Sadly to say they have driven one of the individuals arrested in Arizona, William Rodgers, 40 years old, who runs an environmental infoshop in Flagstaff to committing suicide, at least that is what the Flagstaff jail is currently reporting, but there is an ongoing investigation on this tragic incident.
3. http://www.nal.usda.gov/awic/legislat/pl102346.htm, February 15, 2006.
4. http://judiciary.senate.gov/testimony.cfm?id=1196&wit_id=3460, April 10, 2006.
5. *Animal Liberation Front Guidelines*, Reprinted from the ALF Primer
   - To liberate animals from places of abuse, i.e., laboratories, factory farms, fur farms, etc., and place them in good homes where they may live out their natural lives, free from suffering.
   - To inflict economic damage to those who profit from the misery and exploitation of animals.
   - To reveal the horror and atrocities committed against animals behind locked doors, by performing nonviolent direct actions and liberations.
   - To take all necessary precautions against harming any animal, human and non-human.
   - Any group of people who are vegetarians or vegans and who carry out actions according to ALF guidelines have the right to regard themselves as part of the ALF.
6. See Pattrice Jones,' *Mothers with Monkeywrenches: Feminist Imperatives and the ALF*, in *Terrorists or Freedom Fighters? Reflections on the Liberation of Animals* (Best & Nocella, 2004).
7. http://animalliberationfront.com/Philosophy/BringItToTheTable.htm, February 16, 2007.
8. See Peter Singer, *Animal Liberation.*
9. *Animal Liberation*, pp. xii–xiii. Singer, of course, is a utilitarian, and so his main line of reasoning against ALF tactics logically would be that they could have negative consequences for the movement; thus he might push the pragmatic objection.
10. "Our aim is to destroy property and force laboratories to close—publicity is neither here nor there." Interview with a former ALF activist in "Terrorists or Altruists?" *New Internationalist*, Issue 215, January 1991.
11. See the ALF Primer, www.animalliberation.net.
12. Ledwith, M. (2001). Community work as critical pedagogy: Re-envisioning Freire and Gramsci. *Community Development Journal, 36*(3), 181.
13. See www.richardkahn.org, March 4, 2007.

## *References*

Abu-Jamal, M. (2000). *All things censored.* Ed. N. Hanrahan. New York, NY: Seven Stories Press.

Best, S., & Kahn, R. (2004). Trial by fire: The SHAC7, globalization, and the future of democracy. *Animal Liberation Philosophy and Policy Journal, 2*(2), 1–36.

Best, S., McLaren, P. & Nocella, A. II. (2007). Revolutionary peacemaking: Using a critical pedagogy approach for peacemaking with "terrorists." *Journal for Critical Education Policy Studies 5*(2). 1–37.

Best, S., & Nocella, A. J., II. (2004). *Terrorists or freedom fighters? Reflections on the liberation of animals.* New York, NY: Lantern Books.

Best, S., & Nocella, A. J., II. (2006). *Igniting a revolution: Voices in defense of the Earth.* Oakland, CA: AK Press.

Churchill, W. (2003). Agents of repression: Withstanding the test of time. *Social Justice, 30*(2), 44–50.

Churchill, W., & Vander Wall, J. (2002a). *The COINTELPRO papers: Documents from the FBI's secret war against dissent in the United States.* Cambridge, MA: South End Press.

Churchill, W., & Vander Wall, J. (2002b). *Agents of repression: The FBI's secret wars against the black panther party and the American Indian Movement.* Cambridge, MA: South End Press.

Derrida, J. (2001). *Writing and difference.* New York, NY: Routledge.

Freire, A. M. A., & Macedo, D. (Eds.). (1998). *The Paulo Freire reader.* New York, NY: Continuum Publishing Company.

Freire, P. (1985). *The politics of education: Culture, power, and liberation.* Westport, CT: Bergin and Garvey.

Freire, P. (1997). *Pedagogy of the oppressed.* New York, NY: Continuum Publishing Company.

Gandhi, M. (1993). *Gandhi an autobiography: The story of my experiments with truth.* Boston, MA: Beacon Press.

Giroux, H. (1988). *Teachers as intellectuals: Toward a critical pedagogy of learning.* Granby, MA: Bergin and Garvey.

Giroux, H. (1997). *Pedagogy and the politics of hope: Theory, culture and schooling.* Boulder, CO: Westview Press.

Gramsci, A. (1989). *Selections from the prison notebooks* (10th ed.). New York, NY: International Publishers.

Herman, E., & Chomsky, N. (1988). *Manufacturing consent: The political economy of the mass media.* New York, NY: Pantheon.

hooks, b. (1994). *Teaching to transgress: Education as the practice of freedom.* New York, NY: Routledge.

Irwin, J. (1996). *Empowering ourselves and transforming schools: Educators making a difference.* Albany, NY: State University of New York Press.

Jones, C. E. (1988). Political repression of the Black Panther Party 1966–1971: The case of Oakland bay area. *Journal of Black Studies, 18*(4), 415–435.

Jones, C. E. (1998). *The Black Panther Party reconsidered.* Baltimore, MD: Black Classic Press.

Kahn, R. (2003). Towards ecopedagogy: Weaving a broad-based pedagogy of liberation for animals, nature, and the oppressed people of the Earth. *Animal Liberation Philosophy and Policy Journal, 1*(1), 1–18

Kanpol, B. (1999). *Critical pedagogy: An introduction* (2nd ed.). Westport, CT: Bergin & Garvey.

Kincheloe, J. (2004). *Critical pedagogy primer.* New York, NY: Peter Lang Publishing.

King, M., Jr. (2001).*The autobiography of Martin Luther King, Jr.* New York, NY: Warner Books.

Lather, P. (2001). Ten years later, yet again: Critical pedagogy and it complicities. In K. Weiler (Ed.), *Feminist engagements: Reading, resisting, and revisioning male theorists in education and cultural studies.* New York, NY: Routledge.

Loewen, J. (1995). *Lies my teacher told me: Everything your American history textbook got wrong.* New York, NY: Touchstone.

McLaren, P. (2000). *Che Guevara, Paulo Freire, and the pedagogy of revolution.* Lanham, MD: Rowman & Littlefield Publishers.

McLaren, P. (2006). *Life in schools: An introduction to critical pedagogy in the foundations of education.* Boston, MA: Allyn & Bacon.

Parenti, M. (1993). *Inventing reality: The politics of the news media* (2nd ed.). New York, NY: St. Martin's Press.

Peltier, L. (1999). *Prison writings: My life is my sun dance.* New York, NY: St. Martin's Press.

Shakur, A. (1987). *Assata: An autobiography.* New York, NY: Lawrence Hill Books.

Shor, I. (1992). *Empowering education: Critical teaching for social change.* Chicago, IL: The University of Chicago Press.

West, J (1992). *The Quaker reader.* Philadelphia, PA: Pendle Hill Publications.

Zinn, H. (1999). *A people's history of the United States: 1492–present.* New York, NY: Perennial Classics.

Zinn, H., & Arnove, A. (2004). *Voices of a people's history of the United States.* New York, NY: Seven Stories Press.

# 2. *Beyond Human, Beyond Words: Anthropocentrism, Critical Pedagogy, and the Poststructuralist Turn**

ANNE C. BELL AND CONSTANCE L. RUSSELL

> For the largest part of our species' existence, humans have negotiated relationships with every aspect of the sensuous surroundings, exchanging possibilities with every flapping form, with each textured surface and shivering entity that we happened to focus upon. All could speak, articulating in gesture and whistle and sigh a shifting web of meanings that we felt on our skin or inhaled through our nostrils or focused with our listening ears, and to which we replied... (Abram, 1996, p. ix).

> ...it is language in the form of conflicting discourses which constitutes us as conscious thinking subjects and enables us to give meaning to the world and to act to transform it (Weedon, 1987, p. 32).

Poststructuralist theories have provided a discursive framework through which to critique and contest many of the key tenets of humanism. In drawing attention to the cultural and historical specificity of all human knowledge, they have been used to disrupt assumptions about objectivity, the unified subject, and the universality of human experience, and thereby to expose the classist, racist, sexist, and heterosexist underpinnings of Western humanist thought. For this reason, poststructuralism offers promising theoretical perspectives for educators who wish to challenge cultural representations and structures that give rise to inequities.

Although we acknowledge the important contribution of poststructuralism to analyses of oppression, privilege, and power in education, we believe that educators must continue to probe its limitations and implications. Accordingly, we consider here how poststructuralism, as it is taken up within

* This chapter is reproduced with permission from the *Canadian Journal of Education* (published in 2000, Vol. 25, No. 3, pp. 188–203).

critical pedagogy, tends to reinforce rather than subvert deep-seated humanist assumptions about humans and nature by taking for granted the "borders" (as in Giroux, 1991) that define nature as the devalued Other. We ask what meanings and voices have been preempted by the virtually exclusive focus on humans and human language in a human-centered epistemological framework. At the same time, we discuss how relationships between language, communication, and meaningful experience are being conceptualized outside the field of critical pedagogy (in some cases from a poststructuralist perspective) to call into question these very assumptions. Although we concentrate primarily on societal narratives that shape understandings of human and nature, we also touch on two related issues of language: the "forgetting" of nonverbal, somatic experience and the misplaced presumption of human superiority based on linguistic capabilities. In so doing, our intention is to deal constructively with some of the anthropocentric blind spots within critical pedagogy generally and within poststructuralist approaches to critical pedagogy in particular. We hope to illuminate places where these streams of thought and practice move in directions compatible with our own aspirations as educators.

To borrow from poststructuralism and yet remain within a critical pedagogy framework gives rise, of course, to inevitable tensions. Critical pedagogy continues earlier traditions such as "progressive," "radical," "emancipatory," and "liberation" pedagogies whose root metaphors are distinctly modern (see Bowers, 1993a, pp. 25–26). Poststructuralism, however, brings into play postmodern perspectives and methods of analysis that challenge modernist notions of, for example, freedom, history and progress, rationality, and subjectivity. Nevertheless, poststructuralism has influenced critical theories of education for over a decade, generating fruitful discussions about epistemic certainty, master narratives, stable signifiers, and essentialized identities. As environmental educators, we have found poststructuralism, in concert with the many other theoretical perspectives informing critical pedagogy (e.g., feminism, Marxism, antiracism, Freudian theory, popular education), to be useful in our efforts to come to terms with dominant assumptions about education.

We recognize, furthermore, that poststructuralism as it is taken up within critical pedagogy is only one manifestation of poststructuralist approaches in the human sciences. The term *poststructuralist* applies to a range of (not necessarily coherent) theoretical perspectives. The fact that the term is used differently in Australia, the United States, and Canada further complicates matters (Luke & Luke, 1995, p. 359). Despite important differences, however, forms of poststructuralism share certain assumptions about language, meaning, and subjectivity. A common factor is the analysis of language as "the place where social and political consequences are defined and contested" (Weedon, 1987, pp. 20–21).

## Grounds for Consideration

We come to critical pedagogy with a background in environmental thought and education. Of primary concern and interest to us are relationships among humans and the "more-than-human world" (Abram, 1996), the ways in which those relationships are constituted and prescribed in modern industrial society, and the implications and consequences of those constructs. As a number of scholars and nature advocates have argued, the many manifestations of the current environmental crisis (e.g., species extinction, toxic contamination, ozone depletion, topsoil depletion, climate change, acid rain, deforestation) reflect predominant Western concepts of nature, nature cast as mindless matter, a mere resource to be exploited for human gain (Berman, 1981; Evernden, 1985; Merchant, 1980). An ability to respond adequately to the situation therefore rests, at least in part, on a willingness to critique prevailing discourses about nature and to consider alternative representations (Cronon, 1996; Evernden, 1992; Hayles, 1995). To this end, poststructuralist analysis has been and will continue to be invaluable.

It would be an all-too-common mistake to construe the task at hand as one of interest only to environmentalists. We believe, rather, that disrupting the social scripts that structure and legitimize the human domination of nonhuman nature is fundamental not only to dealing with environmental issues, but also to examining and challenging oppressive social arrangements. The exploitation of nature is not separate from the exploitation of human groups. Ecofeminists and activists for environmental justice have shown that forms of domination are often intimately connected and mutually reinforcing (Bullard, 1993; Gaard, 1997; Lahar, 1993; Sturgeon, 1997). Thus, if critical educators wish to resist various oppressions, part of their project must entail calling into question, among other things, the instrumental exploitive gaze through which we humans distance ourselves from the rest of nature (Carlson, 1995).

For this reason, the various movements against oppression need to be aware of and supportive of each other. In critical pedagogy, however, the exploration of questions of race, gender, class, and sexuality has proceeded so far with little acknowledgement of the systemic links between human oppressions and the domination of nature. The more-than-human world and human relationships to it have been ignored, as if the suffering and exploitation of other beings and the global ecological crisis were somehow irrelevant. Despite the call for attention to voices historically absent from traditional canons and narratives (Sadovnik, 1995, p. 316), nonhuman beings are shrouded in silence. This silence characterizes even the work of writers who call for a rethinking of all culturally positioned essentialisms.

Like other educators influenced by poststructuralism, we agree that there is a need to scrutinize the language we use, the meanings we deploy, and the epistemological frameworks of past eras (Luke & Luke, 1995, p. 378). To treat social categories as stable and unchanging is to reproduce the prevailing relations of power (Britzman, Santiago-Valles, Jiminez-Munes, & Lamash, 1991, p. 89). What would it mean, then, for critical pedagogy to extend this investigation and critique to include taken-for-granted understandings of "human," "animal," and "nature?"

This question is difficult to raise precisely because these understandings are taken for granted. The anthropocentric bias in critical pedagogy manifests itself in silence and in the asides of texts. Since it is not a topic of discussion, it can be difficult to situate a critique of it. Following feminist analyses, we find that examples of anthropocentrism, like examples of gender symbolization, occur "in those places where speakers reveal the assumptions they think they do not need to defend, beliefs they expect to share with their audiences" (Harding, 1986, p. 112).

Take, for example, Freire's (1990) statements about the differences between "Man" and animals. To set up his discussion of praxis and the importance of "naming" the world, he outlines what he assumes to be shared, commonsensical beliefs about humans and other animals. He defines the boundaries of human membership according to a sharp, hierarchical dichotomy that establishes human superiority. Humans alone, he reminds us, are aware and self-conscious beings who can act to fulfill the objectives they set for themselves. Humans alone are able to infuse the world with their creative presence, to overcome situations that limit them, and thus to demonstrate a "decisive attitude towards the world" (p. 90). Freire (1990, pp. 87–91) represents other animals in terms of their lack of such traits. They are doomed to passively accept the given, their lives "totally determined" because their decisions belong not to themselves but to their species. Thus, whereas humans inhabit a "world" which they create and transform and from which they can separate themselves, for animals there is only habitat, a mere physical space to which they are "organically bound."

To accept Freire's assumptions is to believe that humans are animals only in a nominal sense. We are different not in degree but in kind, and though we might recognize that other animals have distinct qualities, we as humans are somehow more unique. We have the edge over other creatures because we are able to rise above monotonous, species-determined biological existence. Change in the service of human freedom is seen to be our primary agenda. Humans are thus cast as active agents whose very essence is to transform the world—as if somehow acceptance, appreciation, wonder, and reverence were beyond the pale.

This discursive frame of reference is characteristic of critical pedagogy. The human/animal opposition upon which it rests is taken for granted, its cultural and historical specificity not acknowledged. And therein lies the problem. Like other social constructions, this one derives its persuasiveness from its "seeming facticity and from the deep investments individuals and communities have in setting themselves off from others" (Britzman et al., 1991, p. 91). This becomes the normal way of seeing the world, and like other discourses of normalcy, it limits possibilities of taking up and confronting inequities (see Britzman, 1995). The primacy of the human enterprise is simply not questioned.

Precisely how an anthropocentric pedagogy might exacerbate the environmental crisis has not received much consideration in the literature of critical pedagogy, especially in North America. Although there may be passing reference to planetary destruction, there is seldom mention of the relationship between education and the domination of nature, let alone any sustained exploration of the links between the domination of nature and other social injustices. Concerns about the nonhuman are relegated to environmental education. And since environmental education, in turn, remains peripheral to the core curriculum (Gough, 1997; Russell, Bell, & Fawcett, 2000), anthropocentrism passes unchallenged.[1]

## *Roots of a Critique*

Bowers (1993a, 1993b) has identified a number of root metaphors or "analogs" in critical pedagogy that reinforce the problem of anthropocentric thinking. These include the notion of change as inherently progressive, faith in the power of rational thought, and an understanding of individuals as "potentially free, voluntaristic entities who will take responsibility for creating themselves when freed from societal forms of oppression" (1993a, pp. 25–26). Such assumptions, argues Bowers, are part of the Enlightenment legacy on which critical pedagogy, and indeed liberal education generally, is based. In other words, they are culturally specific and stem from a period in Western history when the modern industrial world view was beginning to take shape.

To be fair, Bowers understates the extent to which these assumptions are being questioned within critical pedagogy (e.g., Giroux, 1995; Peters, 1995; Shapiro, 1994; Weiler & Mitchell, 1992, pp. 1, 5). Nevertheless, his main point is well taken: proponents of critical pedagogy have yet to confront the ecological consequences of an educational process that reinforces beliefs and practices formed when unlimited economic expansion and social progress seemed promised (Bowers, 1993b, p. 3). What happens when the expansion of human possibilities is equated with the possibilities of consumption? How is educating for freedom predicated on the exploitation of the nonhuman? Such queries push against taken-for-granted understandings of human, nature, self, and community and thus bring into focus the underlying tension between "freedom" as it is constituted within critical pedagogy and the limits that emerge through consideration of humans' interdependence with the more-than-human world.

This tension is symptomatic of anthropocentrism. Humans are assumed to be free agents separate from and pitted against the rest of nature, our fulfillment predicated on overcoming material constraints. This assumption of human difference and superiority, central to Western thought since Aristotle (Abram, 1996, p. 77), has long been used to justify the exploitation of nature by and for humankind (Evernden, 1992, p. 96). It has also been used to justify the exploitation of human groups (e.g., women, Blacks, queers, Indigenous peoples) deemed to be closer to nature—that is, animalistic, irrational, savage, or uncivilized (Gaard, 1997; Haraway, 1989, p. 30; Selby, 1995, pp. 17–20; Spiegel, 1988).

This "organic apartheid" (Evernden, 1992, p. 119) is bolstered by the belief that language is an exclusively human property that elevates mere biological existence to meaningful, social existence. Understood in this way, language undermines our embodied sense of interdependence with a more-than-human world. Rather than being a point of entry into the webs of communication all around us, language becomes a medium through which we set ourselves apart and above.

This view of language is deeply embedded in the conceptual framework of critical pedagogy, including poststructuralist approaches. So too is the human/nature dichotomy upon which it rests. When writers assume that "it is language that enables us to think, speak and give meaning to the world around us," that "meaning and consciousness do not exist outside language" (Weedon, 1987, p. 32), and that "subjectivity is constructed by and in language" (Luke & Luke, 1995, p. 378), then their transformative projects are encoded so as to exclude any consideration of the nonhuman. Such assumptions effectively remove all subjects from nature. As Evernden (1992) puts it, "if subjectivity, willing, valuation, and meaning are securely lodged in the domain of humanity, the possibility of encountering anything more than material objects in nature is nil" (p. 108).

What is forgotten? What is erased when the real is equated with a proliferating culture of commodified signs (see Luke & Luke, 1995, on Baudrillard)? To begin, we forget that we humans are surrounded by an astonishing diversity of life forms. We no longer perceive or give expression to a world in which everything has intelligence, personality, and voice. Polyphonous echoes are reduced to homophony, a term Kane (1994) uses to denote "the reduced sound of human language when it is used under the assumption that speech is something belonging only to human beings" (p. 192). We forget too what Abram (1996) describes as the gestural, somatic dimension of language, its sensory and physical resonance that we share with all expressive bodies (p. 80).

The vast forgetting to which these scholars allude is a culturally and historically specific phenomenon. In Western culture, explains Evernden (1992), it is to the Renaissance that we owe the modern conceptualization of nature from which all human qualities, including linguistic expression, have been segregated and dismissed as "projection." Once scoured of any normative content assigned to humanity, nature is strictly constrained, knowable, and ours to interrogate (pp. 28, 39–40, 48). It is objectified as a "thing," whereas any status as agent or social being is reserved for humans (Haraway, 1988, p. 592).

The language best suited to this cleansing of nature is that of the natural sciences. Scientific accounts, written in language "exclusively descriptive and avowedly neutral" (Evernden, 1992, p. 85), are widely regarded as factual and unbiased and thus are granted a privileged role in naming nature. As Haraway (1986) explains:

> A scientist "names" nature in written, public documents, which are endowed with the special, institutionally enforced quality of being perceived as objective and applicable beyond the cultures of the people who wrote those documents. (p. 79)

According to Haraway (1986), the aesthetic of realism that underlies the truth claims of the natural sciences means many practitioners tend to see themselves not as interpreters but "as discoverers moving from description to causal explanation" (p. 89).

Haraway's analysis reminds us that poststructuralism can and should be used to call into question the universal legitimacy of science insofar as it is used to explicate not only the human domain but also the natural sciences. This questioning almost never takes place. Whereas accusations of reductionism have been leveled at the biobehavioral sciences when focused on humans (e.g., explaining behavior solely in genetic terms), rarely are these accusations made against similar studies on nonhumans (Noske, 1997, p. 83). The reason, presumably, is that the sorts of questions that could be raised about how culture, class, race, and gender shape knowledge about human experience do not pertain to truth claims about the nonhuman. Humans alone are understood to have histories open to interpretation. Everything else is matter for measurement and prediction, physical stuff that can be described and classified once and for all.

To move beyond such taken-for-granted notions of human and nature, Evernden and Haraway suggest, we must admit into the conversation some "non-commonsensical insights" and some "unsettling possibilities" (Evernden, 1992, p. 102; Haraway, 1988, p. 593, respectively). Haraway (1992) writes of "otherworldly conversations," a metaphor helpful in pointing to the possibility of conversants in a discourse in which all of the actors are not "us" (p. 84). To this end, we consider a few promising reconceptualizations of what might constitute language, agency, and meaningful existence beyond the human realm.

### *Otherwordly Conversations*

The human/nature dichotomy is not a frame of reference common to all cultures, and although it prevails today in Western societies, even here there are and always have been alternative ways of understanding and giving expression to a more-than-human world. These can be found, for example, in myth (Kane, 1994, p. 14), poetic expression, certain branches of philosophy and environmental thought, natural history, and children's literature and films (Wilson, 1991, pp. 128–139, 154).

Even within the natural sciences, voices attest to the meaningful existence of nonhuman beings as subjects (McVay, 1993). In animal behavior research, for instance, numerous studies have challenged the assertion of human superiority based on a narrow definition of language that excludes nonhuman communication. Chimpanzee Washoe and orangutan Chantek use American Sign Language, and other primates, like bonobo Kanzi, are fluent in symbolic language, thereby altering the boundaries commonly drawn between language and mere communication (Gardner, Gardner, & Canfort, 1989; Miles, 1994; Savage-Rumbaugh, Shanker, & Taylor, 1998). And though the bilingual great apes may exhibit language patterns the most similar to those of humans, there are many examples of sophisticated communication in other animals, including mammals, birds, and insects (Griffin, 1992).

Meeting the criteria of language implies, of course, that these studies compare and judge other animals against a human yardstick. In other words, a hierarchical divide is still assumed, although its position may shift somewhat to include, on humanity's side, some of the "higher" animals.

For a more radical reframing, one that seeks to acknowledge all life forms as subjects of significance, let us turn to the work of philosopher David Abram. Drawing from phenomenologist Maurice Merleau-Ponty, Abram (1996) argues that all sensing bodies are active, open forms constantly adjusting to a world that is itself continually shifting (p. 49). To demonstrate how all beings incessantly improvise their relations to other things he describes the spontaneous creativity of a spider:

Consider a spider weaving its web, for instance, and the assumption still held by many scientists that the behavior of such a diminutive creature is thoroughly "programmed in hers genes." Certainly, the spider has received a rich genetic inheritance from its parents and predecessors. Whatever "instructions," however, are enfolded within the living genome, they can hardly predict the specifics of the microterrain within which the spider may find herself at any particular moment. They could hardly have determined in advance the exact distances between the cave wall and the branch that the spider is now employing as an anchorage point for her current web, or the exact strength of the monsoon rains that make web-spinning a bit more difficult on this evening. And so the genome could not explicitly have commanded the order of every flexion and extension of her various limbs as she weaves this web into its place. However complex are the inherited "programs," patterns, or predispositions, they must still be adapted to the immediate situation in which the spider finds herself. However determinate one's genetic inheritance, it must still, as it were, be woven into the present, an activity that necessarily involves both a receptivity to the specific shapes and textures of that present and a spontaneous creativity in adjusting oneself (and one's inheritance) to those contours (Abram, 1996, p. 50).

An equally illuminating insect story, intended to evoke, once again, the subjective world of a nonhuman being, is found in Evernden's *The Natural Alien* (1985, pp. 79–80). Borrowing from the work of biologist Jakob von Uexkull, Evernden invites readers "to imagine that we are walking through a meadow and that we discern 'a soap bubble around each creature to represent its own world, filled with the perceptions which it alone knows'" (p. 79). He then attempts to describe what might be the world of a wood tick. The wood tick, he explains, is literally and figuratively blind to the world as we know it. What we readily perceive about our environment would be unknown, unknowable, and irrelevant to her. Her world is composed of three elements: light, sweat, and heat. These are all that she needs to complete her life cycle. Light will lead her to the top of a bush, where she will cling (for as long as 18 years!) until the smell of sweat alerts her to a passing animal. She will then drop, and if she lands on a warm animal, she will indulge in a blood meal, fall to the ground, lay her eggs, and die.

Like Abram, Evernden (1985) challenges commonplace, mechanistic assumptions that reduce other life forms to programmed automatons and intimates instead a meaningful life-world completely unlike and outside our own:

> To speak of reflexes and instincts is to obscure the essential point that the tick's world is a world, every bit as valid and adequate as our own. There is a subject, and like all subjects she has her world... The tick is able to occupy a world that is perceptually meaningful to she. Out of the thousands or millions of kinds of information that might be had, the tick sees only what is of significance to she. The world is tailored to the animal; they are entirely complementary... This is quite a different view of existence from our usual one in which the animal is simply an exploiter of certain natural resources. We are not talking just about observable interactions between subjects and objects but rather about a very complete interrelation of self and world, so complete that the world could serve as a definition of the self. Without the tick there is no tick-world, no tick-space, no tick-time—no tick-reality (pp. 80–81).

Evernden's remarks are significant for the possibilities they open up in our understanding both of the nonhuman and of ourselves. On one hand, they contest the limited notion that awareness is a specifically human attribute. On the other, they remind us that we humans too have bodies that respond to light, sweat, and heat; we too know the world through our bodies in a way that is not entirely dependent upon language; and this bodily knowledge plays an important role in defining our world and giving meaning to it.

## *Shared Conversations*

In challenging anthropocentrism, the two of us find cause for hope in the fact that our critique can be seen as compatible with the work of many proponents of critical pedagogy. Specifically, attention to local contexts, lived relationships, and embodied learning within critical pedagogy matches similar considerations within environmental thought and education. The poststructuralist emphasis on societal narratives and language practices, already well developed in critical pedagogy, is likewise being taken up by environmental scholars and educators. What strikes us as most auspicious, then, is the potential for shared conversations, with insights from one field sparking unasked questions and opening up unexplored pathways for another.

For instance, carrying forward the concerns and convictions of Dewey (1938/1963) and the progressive education movement, theorists of critical pedagogy have written extensively about the disjuncture between the kinds of environments and interactions necessary for active and transformative learning and the social relations we enter into through academic training (e.g., McKenna, 1991). They recommend practices situated in students' cultures (e.g., Shor, 1992, p. 44) and in the particular communities, schools, and other social groups of which students are a part (e.g., Walsh, 1991, p. 99).

In so doing, they stress the importance of relationships, contexts, and local histories in defining who we are, calling into question the individualistic and universalistic narratives that shape curriculum and schooling generally (e.g., Giroux, 1991, p. 24; Weiler & Mitchell, 1992, pp. 1, 5).

So far, however, such queries in critical pedagogy have been limited by their neglect of the ecological contexts of which students are a part and of relationships extending beyond the human sphere. The gravity of this oversight is brought sharply into focus by writers interested in environmental thought, particularly in the cultural and historical dimensions of the environmental crisis. For example, Nelson (1993) contends that our inability to acknowledge our human embeddedness in nature results in our failure to understand what sustains us. We become inattentive to our very real dependence on others and to the ways our actions affect them. Educators, therefore, would do well to draw on the literature of environmental thought in order to come to grips with the misguided sense of independence, premised on freedom from nature, that informs such notions as "empowerment."

Further calls for educational practices situated in the life-worlds of students go hand in hand with critiques of disembodied approaches to education. In both cases, critical pedagogy challenges the liberal notion of education whose sole aim is the development of the individual, rational mind (Giroux, 1991, p. 24; McKenna, 1991, p. 121; Shapiro, 1994). Theorists draw attention to the importance of nonverbal discourse (e.g., Lewis & Simon, 1986, p. 465) and to the somatic character of learning (e.g., Shapiro, 1994, p. 67), both overshadowed by the intellectual authority long granted to rationality and science (Giroux, 1995; Peters, 1995; S. Taylor, 1991). Describing an "emerging discourse of the body" that looks at how bodies are represented and inserted into the social order, S. Taylor (1991) cites as examples the work of Peter McLaren, Michelle Fine, and Philip Corrigan.

A complementary vein of enquiry is being pursued by environmental researchers and educators critical of the privileging of science and abstract thinking in education. They understand learning to be mediated not only through our minds but also through our bodies. Seeking to acknowledge and create space for sensual, emotional, tacit, and communal knowledge, they advocate approaches to education grounded in, for example, nature experience and environmental practice (Bell, 1997; Brody, 1997; Weston, 1996). Thus, whereas both critical pedagogy and environmental education offer a critique of disembodied thought, one draws attention to the ways in which the body is situated in culture (Shapiro, 1994) and to "the social construction of bodies as they are constituted within discourses of race, class, gender, age and other forms of oppression" (S. Taylor, 1991, p. 61). The other emphasizes and celebrates our embodied relatedness to the more-than-human world and to the myriad life forms of which it is comprised (Payne, 1997; Russell & Bell, 1996). Given their different foci, each stream of enquiry stands to be enriched by a sharing of insights.

Finally, with regard to the poststructuralist turn in educational theory, ongoing investigations stand to greatly enhance a revisioning of environmental education. A growing number of environmental educators question the empirical–analytical tradition and its focus on technical and behavioral aspects of curriculum (Gough, 1997; Robottom, 1991). Advocating more interpretive, critical approaches, these educators contest the discursive frameworks (e.g., positivism, empiricism, rationalism) that mask the values, beliefs, and assumptions underlying information, and thus the cultural and political dimensions of the problems being considered (Gough, 1997; Huckle, 1999; Lousley, 1999). Teaching about ecological processes and environmental hazards in a supposedly objective and rational manner is understood to belie the fact that knowledge is socially constructed and therefore partial (Gough, 1997; Robertson, 1994; Robottom, 1991; Stevenson, 1993).

Gough (1999) explicitly goes beyond critical approaches to advocate poststructuralist positions in environmental education. He asks science and environmental educators to adopt skepticism toward metanarratives, an attitude that characterizes poststructuralist discourses. Working from the assumption that science and environmental education are story-telling practices, he suggests that the adequacy of narrative strategies be examined in terms of how they represent and render problematic "human transactions with the phenomenal world" (Gough, 1993, p. 607). Narrative strategies, he asserts, should not create an illusion of neutrality, objectivity, and anonymity, but rather draw attention to our kinship with nature and to "the personal participation of the knower in all acts of understanding" (Gough, 1993, p. 621).

We contend, of course, that Gough's proposal should extend beyond the work of science and environmental educators. The societal narratives that legitimize the domination of nature, like those that underlie racism, sexism, classism, heterosexism, and so on, merit everyone's concern. And since the ecological crisis threatens especially those most marginalized and vulnerable (Running-Grass, 1996; D. Taylor, 1996), proponents of critical pedagogy in particular need to come to terms with the human-centered frameworks that structure their endeavors.

No doubt poststructuralist theory will be indispensable in this regard. Nevertheless, anthropocentric assumptions about language, meaning, and agency will need to be revisited. In the meantime, perhaps we can ponder the spontaneous creativity of spiders and the life-worlds of wood ticks. Such wondrous possibilities should cause even the most committed of humanists to pause for a moment at least.

## Note

1. Although the term *anthropocentrism* brings into play a hierarchical dualism that can mask the complexity of human relationships with other forms of life and deemphasize the permeability of human/nonhuman borders, we find the word helpful, if not indispensable, in naming and resisting a way of being in the world that "places humanity and human interests at the center of value" (Katz, 1997, p. 122).

## References

Abram, D. (1996). *The spell of the sensuous: Perception and language in a more-than-human world*. New York, NY: Pantheon Books.

Bell, A. (1997). Natural history from a learner's perspective. *Canadian Journal of Environmental Education*, *2*, 132–144.

Berman, M. (1981). *The reenchantment of the world*. Ithaca, NY: Cornell University Press.

Bowers, C. A. (1993a). *Critical essays on education, modernity and the recovery of the ecological imperative*. New York, NY: Teachers College Press.

Bowers, C. A. (1993b). *Education, cultural myths, and the ecological crisis: Toward deep changes*. Albany, NY: State University of New York Press.

Britzman, D. (1995). Is there a queer pedagogy? Or, stop reading straight. *Educational Theory*, *45*(2), 151–165.

Britzman, D., Santiago-Valles, K., Jiminez-Munes, G., & Lamash, L. (1991). Dusting off the erasures: Race, gender and the problem of pedagogy. *Education and Society*, *9*(1/2), 83–99.

Brody, M. (1997). Descending the watershed: Rethinking the place of curriculum. *Canadian Journal of Environmental Education*, *2*, 114–131.

Bullard, R. D. (1993). Anatomy of environmental racism and the environmental justice movement. In R. D. Bullard (Ed.), *Confronting environmental racism: Voices from the grassroots* (pp. 15–39). Boston, MA: South End Press.

Carlson, D. (1995). Making progress: Progressive education in the postmodern. *Educational Theory*, *45*, 337–357.

Cronon, W. (1996). Introduction: In search of nature. In W. Cronon (Ed.), *Uncommon ground: Rethinking the human place in nature* (pp. 23–56). New York, NY: W.W. Norton & Co.

Dewey, J. (1963). *Experience and education*. New York, NY: Macmillan. (Original work published 1938).

Evernden, N. (1985). *The natural alien: Humankind and environment*. Toronto, Canada: University of Toronto Press.

Evernden, N. (1992). *The social creation of nature.* Baltimore, MD: Johns Hopkins University Press.

Freire, P. (1990). *Pedagogy of the oppressed.* New York, NY: Continuum.

Gaard, G. (1997). Toward a queer ecofeminism. *Hypatia, 12*(1), 114–137.

Gardner, R. A., Gardner, B. T., & Canfort, T. E. (Eds.). (1989). *Teaching sign language to chimpanzees.* Albany, NY: State University of New York Press.

Giroux, H. A. (1991). Border pedagogy and the politics of postmodernism. *Education and Society, 9*(1), 23–38.

Giroux, H. A. (1995). Series foreword. In M. Peters (Ed.), *Education and the post-modern condition* (pp. ix–xvii). Westport, CT: Bergin and Garvey.

Gough, A. (1997). *Education and the environment: Policy, trends and the problems of marginalisation.* Melbourne: Australian Council for Educational Research.

Gough, N. (1993). Environmental education, narrative complexity and postmodern science/fiction. *International Journal of Science Education, 15*, 607–625.

Gough, N. (1999). Rethinking the subject: (de)constructing human agency in environmental education research. *Environmental Education Research, 5*(1), 35–48.

Griffin, D. R. (1992). *Animal minds.* Chicago, IL: University of Chicago Press.

Haraway, D. (1986). Primatology is politics by other means. In R. Bleier (Ed.), *Feminist approaches to science* (pp. 77–117). New York, NY: Pergamon Press.

Haraway, D. (1988). Situated knowledges: The science question in feminism and privileges of partial perspectives. *Feminist Studies, 14*(3), 575–599.

Haraway, D. (1989). *Primate visions: Gender, race and nature in the world of modern science.* New York, NY: Routledge.

Haraway, D. (1992). Otherworldly conversations: Terran topics; local terms. *Science as Culture, 3*(14), 64–98.

Harding, S. (1986). *The science question in feminism.* Ithaca, NY: Cornell University Press.

Hayles, N. K. (1995). Searching for common ground. In M. Soule & G. Lease (Eds.), *Reinventing nature? Responses to postmodern deconstruction* (pp. 47–63). Washington, DC: Island Press.

Huckle, J. (1999). Locating environmental education between modern capitalism and postmodern socialism: A reply to Lucie Sauvé. *Canadian Journal of Environ-mental Education, 4*, 36–45.

Kane, S. (1994). *The wisdom of the mythtellers.* Peterborough, ON: Broadview Press.

Katz, E. (1997). *Nature as subject: Human obligation and natural community.* Lanham, MD: Rowman & Littlefield.

Lahar, S. (1993). Roots: Rejoining natural and social history. In G. Gaard (Ed.), *Ecofeminism: Women, animals, nature* (pp. 91–117). Philadelphia, PA: Temple University Press.

Lewis, M., & Simon, R. (1986). "A discourse not intended for her": Teaching and learning within patriarchy. *Harvard Educational Review, 56*, 457–472.

Lousley, C. (1999). (De)politicizing the environment club: Environmental discourses and the culture of schooling. *Environmental Education Research, 5*(3), 293–304.

Luke, C., & Luke, A. (1995). Just naming? Educational discourses and the politics of identity. In W. T. Pink & G. W. Noblit (Eds.), *Continuity and contradiction: The futures of the sociology of education* (pp. 357–380). Cresskill, NJ: Hampton Press.

McKenna, K. (1991). Subjects of discourse: Learning the language that counts. In H. Bannerji, L. Carty, K. Dehli, S. Heald, & K. McKenna (Eds.), *Unsettling relations: The university as a site of feminist struggles* (pp. 109–128). Toronto: The Women's Press.

McVay, S. (1993). Prelude: A Siamese connexion with a plurality of other mortals. In E. O. Wilson & S. R. Kellert (Eds.), *The biophilia hypothesis* (pp. 3–19). Washington, DC: Island Press.

Merchant, C. (1980). *The death of nature: Women, ecology, and the scientific revolution.* New York, NY: Harper and Row.

Miles, H. L. W. (1994). ME CHANTEK: The development of self-awareness in a signing orangutan. In S. T. Parker, R. W. Mitchell, & M. L. Boccia (Eds.), *Self-awareness in animals and humans: Developmental perspectives* (pp. 254–272). New York, NY: Cambridge University Press.

Nelson, R. (1993). Searching for the lost arrow: Physical and spiritual ecology in the hunter's world. In S. R. Kellert & E. O. Wilson (Eds.), *Biophilia hypothesis* (pp. 201–228). Washington, DC: Island Press.

Noske, B. (1997). *Beyond boundaries: Humans and animals.* Montreal: Black Rose Books.

Payne, P. (1997). Embodiment and environmental education. *Environmental Education Research*, 3(2), 133–153.

Peters, M. (1995). Introduction: Lyotard, education and the postmodern condition. In M. Peters (Ed.), *Education and the postmodern condition* (pp. xxiii–xliii). Westport, CT: Bergin and Garvey.

Robertson, A. (1994). Toward constructivist research in environmental education. *Journal of Environmental Education, 25*(2), 21–31.

Robottom, I. (1991). Technocratic environmental education: A critique and some alternatives. *Journal of Experiential Education, 14*(1), 20–26.

Running-Grass. (1996). The four streams of multicultural environmental education. *Race, Poverty and the Environment, 6*(2/3), 1–2.

Russell, C. L., & Bell, A. C. (1996). A politicized ethic of care: Environmental education from an ecofeminist perspective. In K. Warren (Ed.), *Women's voices in experiential education* (pp. 172–181). Dubuque, IO: Kendall/Hunt Publishing.

Russell, C. L, Bell, A. C., & Fawcett, L. (2000). Navigating the waters of Canadian environmental education. In T. Goldstein & D. Selby (Eds.), *Weaving connections: Education for peace, social and environmental justice* (pp. 196–217). Toronto: Sumach Press.

Sadovnik, A. R. (1995). Postmodernism in the sociology of education: Closing the rift among theory, practice, and research. In W. T. Pink & G. W. Noblit (Eds.), *Continuity and contradiction: The futures of the sociology of education* (pp. 309–325). Cresskill, NJ: Hampton Press.

Savage-Rumbaugh, S., Shanker, S. G., & Taylor, T. J. (1998). *Apes, language, and the human mind*. New York, NY: Oxford University Press.

Selby, D. (1995). *Earthkind: A teachers' handbook on humane education*. London: Trentham Books Ltd.

Shapiro, S. (1994). Re-membering the body in critical pedagogy. *Education and Society, 12*(1), 61–79.

Shor, I. (1992). *Empowering education: Critical teaching for social change*. Chicago, IL: University of Chicago Press.

Spiegel, M. (1988). *The dreaded comparison: Human and animal slavery*. Philadelphia, PA: New Society Publishers.

Stevenson, R. B. (1993). Becoming compatible: Curriculum and environmental thought. *Journal of Environmental Education, 24*(3), 4–9.

Sturgeon, N. (1997). *Ecofeminist natures: Race, gender, feminist theory, and political action*. New York, NY: Routledge.

Taylor, D. E. (1996). Making multicultural environmental education a reality. *Race, Poverty and the Environment, 6*(2/3), 3–6.

Taylor, S. (1991). Skinned alive: Towards a postmodern pedagogy of the body. *Education and Society, 9*(1), 61–72.

Walsh, C. E. (1991). *Pedagogy and the struggle for voice*. Toronto: OISE Press.

Weedon, C. (1987). *Feminist practice and poststructuralist theory*. London: Blackwell.

Weiler, K., & Mitchell, C. (1992). *What schools can do: Critical pedagogy and practice*. Albany, NY: State University of New York Press.

Weston, A. (1996). Deschooling environmental education. *Canadian Journal of Environmental Education, 1*, 35–46.

Wilson, A. (1991). *The culture of nature: North American landscape from Disney to the Exxon Valdez*. Toronto: Between the Lines.

# *Part II: Pedagogy*

# 3. *Critical Animal Pedagogy: Explorations Toward Reflective Practice*

KARIN GUNNARSSON DINKER AND HELENA PEDERSEN

> Under Empire, the old Institutions are deteriorating one after the other in a cascade of apparatuses. What is happening, and what is the truly imperial mission, is the concerted demolition of each Institution into a multiplicity of apparatuses, into an arborescence of relative and unpredictable norms. The educational system, for example, no longer bothers to present itself as a coherent order. It is now but a hodgepodge of classes, schedules, subjects, buildings, departments, programs, and projects that are so many apparatuses meant to keep bodies immobilized. (Tiqqun 2011, pp. 201–202)

## *Introduction*

Above, Tiqqun (2011) draws our attention to the increasingly fragmented and seemingly haphazard character of the education system in the 21st century. With this brief analysis in mind, our chapter asks what work critical animal pedagogy, as reflective and radical educational practice, can do in such an incoherent terrain?

Drawing on a long tradition of critical theoretical approaches to education (e.g., Giroux, 1981; hooks, 1981; Illich, 1970; McLaren & Kincheloe, 2007), we argue that education is not necessarily an inherent "common good," offered for the benefit of public welfare. Rather, education actively (re)produces injustices and inequalities prevalent in other sectors in society. The education system is also a site of power struggles, since those who have power over education and educational processes (be it the church, the nation state, or the private sector) also are believed to exert an influence over our contemporary as well as future society. Education theorist Clayton Pierce (2013) has argued, for instance, that educational institutions have become an important part of the biopolitical landscape, a terrain where life and its mechanisms are targeted by a range of production and control regimes, forming a new educational anatomy aiming at the dissolution of restrictive barriers to the commodification of life (biocapital). These, and other, biopolitical arrangements of power and domination in education have, specifically in relation to animals, been reframed by Patricia MacCormack (2013) as a *war* between humans and other species.

As we outline elsewhere (Gunnarsson Dinker, forthcoming; Gunnarsson Dinker & Pedersen, 2016; Pedersen, 2010, 2011a), the position of animals in (formal) education has a long history of exploitation as teaching and learning tools, scientific objects, species representatives, and food served in the school canteen. We follow MacCormack's (2013) argument that these (asymmetric) interspecies relations of power and domination are an inbuilt component of education itself, if education, ultimately, is a process of becoming integrated into a profoundly anthropocentric, capitalist-driven, unequal, consumerist society.

What, then, could be the role of critical animal pedagogy in times when education as such is under siege by commercial and other vested interests (see Aronowitz & Giroux, 1986; cf. Biesta & Säfström, 2011)? Following Rowe (2011), we see an important role of critical animal pedagogy as *disturbing*: Disturbing norms, practices, cultures, and structures that render animals endlessly accessible for human use. The widespread anxiety about critical pedagogies in general as "normative"—a heated debate haunting also other educational subfields such as education for sustainable development (Kopnina & Cherniak, 2015)—is, we believe, invalidated by the fact that education, as policy and practice, is *always already* normative and cannot be otherwise. However, this normativity must not be confused with being programmatic, but should continuously provide space for critical reflection: Reflection on how to think outside the box, how to read in new ways, how to eat in new ways, how to teach and learn in new ways. Reflection on how critical educators, students, and animals may jointly create new educational and world-forming conditions (cf. Palmer & Pedersen, 2017).

In this chapter we delineate some thoughts, drawing on previous research in our own and related fields as well as on our own experiences as educators in various contexts, on how to move forward with critical animal pedagogy as a disturbing and disruptive process, a reflective practice, and a countermovement to animal, human, and environmental exploitation. Many examples will be drawn from Karin's ethnographic fieldwork experience with state-funded primary schools in Sweden, one of them a Steiner school following a Waldorf curriculum (with a curricular emphasis on, for instance, practical work and arts), and also from Helena's research and teaching with upper secondary schools and at the university level. (See Linné & Pedersen, 2014 for a more in-depth account of critical animal pedagogy in higher education.)

Each section addresses what we see as a particularly central dimension of critical animal pedagogy and raises a specific pedagogical question or problem which will be used to guide our discussion on pedagogical work with animal and human liberation. Our discussions in this chapter focus primarily on farmed animals. The sections are modeled on our recently published, coauthored book chapter (Gunnarsson Dinker & Pedersen, 2016). In the present chapter, however, we foreground personal pedagogical experiences and reflections (our own and others'), rather than their theoretical underpinnings (covered at length in the previous chapter, where we advocate a "new abolitionist" approach to animals and pedagogy, as developed by MacCormack, 2013). We hope, however, that these two texts, in combination, offer complementary contributions to the emerging scholarly-activist dialogue on how critical animal pedagogies may be conceptualized, developed, and put to reflective practice across educational levels and curricula.

## *The Power of Emotion*

Emotion, as an inherent part of any educational experience, may be appropriated and manipulated for a multitude of purposes, and through a range of technologies, in education (Cole & Stewart 2014; Pedersen 2013, 2015). The meat and dairy lobby typically works with affective promotion materials and events in their outreach to potential consumers, including school teachers and children (Linné & Pedersen, 2016). How can the complexities of affect be addressed in critical animal pedagogy?

As a response to these strategies, we have suggested a critical-analytic approach (Gunnarsson Dinker & Pedersen, 2016), encouraging students to engage critically with the animal industry's messages, along with the societal norms, discourses, and institutions accompanying them, and investigate how they work on, and through, students' own emotions. One critical-analytic approach is to introduce elements of critical media literacy to make visible how affect is exploited by the animal industry. This includes questioning the ways that meat, dairy, and other animal industries work on consumers' affective responses by analyzing "emotionally charged animal imagery" (Gunnarsson Dinker & Pedersen, 2016), including the depiction of the industrial farm as a local and cozy family farm (Cole & Stewart, 2014; Yates, 2005), and the "edutainment" events provided by the industry, such as online games (Cole & Stewart, 2014) or "pasture releases" at dairy farms (Linné & Pedersen, 2016).

In our fieldwork experience, reaching out to schools with materials that aid the development of critical-analytic skills by offering critical accounts of the animal industry's promotional strategies has several problems. First, there is a huge lack of teaching and learning materials in Swedish that question the claims and rhetoric of the industry, at least on the level of compulsory school. Opportunities to introduce critical accounts of how we relate to animals in human society are few. Even so-called humane education materials, such as "REDE" (Malm, 2007) (*Respect, Empathy, Animals, Ethics: A values education material with the aim to develop children's empathic ability and respectful view toward animals and nature*; our translation), developed by Animal Welfare Sweden for ages 6–12, which are designed also to develop children's awareness of their own emotions in productive and creative ways, tend to downplay or even exclude the situation of farmed animals. As a result, the sensitive issue of veganism and vegetarianism is avoided, and the paradoxical message that animals may be both eaten and loved is, although the aim of the material is the opposite, reinforced. This kind of "emotional training" may actually work to keep intact *status quo* in society's relation to animals—an ever-present risk when education does not take place in an environment that fully supports justice and emotional engagement for animals and their liberation.

Presenting critical materials as countermessages to promotion materials from the animal industry may be further complicated by the age group and the nature of the material. For instance, the farm theme is often dealt with in primary school, when the children are considered too young for exposure to how animals are really treated. We suspect this is in line with the widespread idea (Sobel, 1996) in environmental education that children under 10 exposed to crisis scenarios produce only negative affective response. But evidence for any such claim is very weak and new research shows that children do have strategies for coping and acting on environmental information including those exposing the vulnerability of animals (Hussar & Harris, 2009; Myers, 2013). We emphasize again the significance of creating an educational environment that gives full support to students in this and other learning processes.

## *Intersubjectivity and Rebellion*

> "Don't be silly, frog's legs are delicious and just like eating chicken legs!"
> Sharing circle continued but I wasn't listening anymore. I dwelled on frog legs and chicken legs and Mrs. Gladstone's legs.
> Sharing circle was a productive pedagogical exercise that day: I learned how simile worked (although I didn't know the term simile until years later). I learned

> that I lived in a culture that slaughtered and ate animals (and amphibians!) and that my teacher enjoyed this activity and derived a peculiar pleasure in normalizing it for our class (I didn't know the term normalize yet, but I felt and fought its affects). I learned that humans live by sucking the lives out of other animals. I went home and told my parents I would never eat meat again. (Truman, 2016, p. 38)

Truman's (2016) account of how she became vegetarian in kindergarten suggests that it is in the process of identification with animals, rather than in fact-based learning, that children in school may discover their intersubjectivity with nonhumans. Myers (2013) and Cole and Stewart (2014) argue, however, that these possibilities of intersubjectivity are effectively obliterated by the education system. Describing the situation of children in the school system, Taje (2016) writes: "Scholastic power objectifies them, documents and relays knowledge about their 'strengths' and 'weaknesses,' and turns them into 'cases' to be correctly trained for the purposes of adulthood (i.e., work)" (Taje, 2016, p. 4440). In short, if children are oppressed, "not free to follow their own inclinations, desires, or interests" (Taje, 2016, p. 4440), how can we expect them to act on behalf of others? Still—and against all odds, perhaps—students of all ages *do* resist, do empathize, and do act on behalf of others.

> If we wish to move beyond the necrophilic, sadistic, state-sanctioned school system towards ones that respect children and nurture their capacities for freedom, reciprocity, and direct action, *then we need to rebel.* (Taje, 2016, p. 4702; emphasis in original)

In our respective fieldwork we have, for instance, come across students who refuse to participate in dissections though threatened that they will then not be graded, who refuse to eat animal products or eat when these are served, who leave the classroom in protest when teachers give uncritical accounts of farm practices, and who engage in multiple other forms of micro-resistances. We have also encountered students who use their entire education as a platform for animal liberation work (see Pedersen, 2011b, 2012). In her parental role, Karin's rebellion has consisted in encouraging and supporting these acts of resistance among her own children and their peers, a role not always easy to balance against the role of researcher. Teachers are, understandably, often less prone to rebel as it may put their job in jeopardy by conflicting with the social conventions around the ostensibly professional "objectivity" expected of the teacher role (norms strongly rooted also among many students). Other possibilities of resisting anthropocentric and speciesist organization of our affective relations with animals are through activist accounts of shared vulnerability

(Drew, 2014), stories of resistance among animals (Hribal, 2007, 2011), and of joint animal and child resistance in school (Gunnarsson Dinker, forthcoming). Such examples may open new avenues of reflection, inspiration, and possibilities of cross-species intersubjectivity.

### *Conclusion*

The issue of emotion and affect in education is, in many respects, complex, controversial, and risky. While an inherent dimension of the educational experience as such, emotion is a part of asymmetric power relations in education and can also be managed and organized for ostensibly "good," however still coercive, purposes. We stress that the critical animal pedagogy educator must have in-depth insight about the nature of emotion in teaching and learning processes and educational power relations, and act on these insights with respect, reflection, and careful attention to each student, as well as to the dynamics of the collective learning experience. With this in mind, a core component to develop in the critical animal pedagogy classroom or situation should be the ability to read critically the various texts, messages, and actions produced by the animal industries—to make visible how the industry exploits our emotions in order to exploit the animals. Another core component that we have identified is the facilitation of and support for experiences of shared vulnerability. It can take the form of "an embodied way of knowing," a phenomenological, preverbal, and foundational form of knowledge that challenges the hegemonic idea that equates knowledge with rationality (Drew, 2014). Or it can be mediated through literature, film, and other art, activist tales, radicalized humane education materials. All forms can, potentially, become a foundation for developing a sense of justice for animals.

## ***The Animal-Industrial Complex: From Speaking Silences to Animal Voice***

Teaching to disrupt the animal-industrial complex (A-IC)—the increasing interpenetration and network convergence of animal-based agriculture, biotechnologies, livestock genetics corporations, pharmaceutical corporations, and other investment interests in global food systems (Noske, 1997; Twine, 2012)—is one of the most urgent and difficult tasks of critical animal pedagogy. In Sweden, a country celebrating its own progressive "animal friendliness" as a source of national pride (cf. Svärd, 2015), students and teachers tend to resist or even reject a critical inquiry into an exploitative system of production ostensibly taking place "elsewhere." This pedagogical problem, we argue, *is an inherent part of the A-IC itself*, as the A-IC relies on the formal education system—from kindergarten to university level—as sites of reproduction of a plethora of animal oppressive practices, conveniently incorporated in the socialization processes of education (see Linné & Pedersen, 2016; Noske, 1997; Rowe, 2012, 2013). Education must thus be understood as a vital and productive part of the A-IC, rather than as something outside of it—or, to use Pierce's (2013) Foucauldian terminology, educational institutions must be viewed as integrated in a biopolitical landscape largely organized by the A-IC. In Helena's fieldwork in upper secondary school, classroom discussions on animal ethics in the food production system remained in a "safe" and neutralizing sphere, where most problems identified by students were ultimately downplayed, or framed as "deviations" in an otherwise benevolent production system, rather than explored as a structural problem to be critiqued and changed (Pedersen, 2010). Karin found, during her fieldwork with younger schoolchildren, similar versions of "speaking silences" in relation to farmed animals:

> On returning from a farm visit the school children are asked to recall the trip by their teachers. The calves, clearly, have had the most significant impact. Several children recall the calves licking them, chewing their clothes and sucking their fingers. When the teacher asked the children to consider the reasons for these practices one stated, "They [the calves] think the finger is the cow's teat" to which the teacher replied, "Yes the calves miss their mother." While the teacher returned to the calf's behavior several times during the lesson the fact that calves are taken from their mothers is itself never questioned or discussed. (Gunnarsson Dinker, forthcoming)

The teacher cited above is a vegetarian. Vegan or vegetarian educators, guided by real or imagined expectations of their professional role of the "objective" teacher, indeed seem to discipline themselves into performing a kind of self-censorship when it comes to (farmed) animals' rights debates in the classroom. These "speaking silences" then effectively contribute to sustaining the A-IC through the school context: Exploitative practices of animal farming and slaughter are deliberately hidden in education (Rowe, 2012) as are the vast numbers of animals killed every day for human consumption (Nowatzki, 2013).

### *Where Does Food Come From?*

How can "speaking silences" in education be challenged? As critical animal pedagogy educators, we will often find our pedagogical practices taking shape through a repertoire of rebellious acts that may also involve even reformistic material. Animal advocacy organizations can produce excellent material, but if it is considered too "radical" in school it is not allowed to enter. Many organizations thus make their efforts on another level, producing material considered less controversial. Although we believe this trade-off should be criticized, in Karin's experience, so-called humane education material (advocating animal protection rather than liberation) may be used as a "stepping stone" to carve out space for more radical approaches. During fieldwork, Karin was present at lessons falling under the humane education umbrella and in the end was given the possibility to present a lesson about an "animal sanctuary" as an alternative kind of farm. During this lesson children themselves came up with questions such as "Do we really need to eat animals at all?" Opening up the question of whether we really need animal products at all can be a very productive entrance point into unsettling the whole basis of the A-IC—particularly, we believe, when working with students of younger ages. The fact that we might not need any of those products is, in education and beyond, a carefully guarded secret, a "speaking silence" and an emperor in no clothes that needs to come out in the open.

> When slaughter was eventually touched upon in a primary school farm theme lesson it made it obvious to me that such accounts need to be taken out of the box, added to, since the whole idea that we need to kill animals for our own used to uphold meat normativity. (Gunnarsson Dinker, forthcoming)

Showing children "where food comes from" is a celebrated pedagogical idea that is frequently co-opted as promotion strategy by the animal industries (see Linné & Pedersen, 2016), reinforcing values and myths that serve their purposes: we have to eat meat, farming is environmentally friendly and historically, humans have always done so. The image of the small-scale, organic family farm is a frequently recurring trope in this ideological work, conveyed through children's literature, film, and other forms of popular culture and media. Embedded in childhood, these images are very powerful (Yates, 2013), gloss over and make reasonable (Cole & Stewart, 2014) the reality of confinement, deprivation, separations, mutilations, and slaughter that also so-called "organic" animals endure. Adult society *knows*, however, what is at stake: In an account of her childhood on her family's farm, Trachsel (2017) tells us that when it was time for the pigs to be taken away to slaughter, her mother would take the children into the bathroom, close the door, and turn on the water in the bath tub to drown out the cries of the pigs as they were forced into the truck. We argue that (age-appropriate) exposure to what really happens to the supposedly happy animals in the farmyard, during transport and slaughter, *is a right of the child*, both in terms of acquiring the critical literacy skills required to critically interpret children's global commodity cultures such as computer games, literature, and film, and in children's roles as consumers of other commodities such as food and clothes. The ability to analyze speciesist messages that reinforce and maintain the commodity status of animals is a basic necessity for such an understanding. For this purpose, we find Stewart and Cole's (2009) identification of hidden messages helpful in the analyses of popular culture material. Applied to pedagogical practice, we trust that this approach could be used as a basis for dialogue with the children themselves. This requires that, in contrast to the superficiality of standardized learning, authentic critical and deep reflection is allowed and encouraged in the classroom. While some might argue that deep reflection might be difficult to endorse among younger students who (presumably) have yet to achieve the proper level of maturity, we propose that it is not. The following example will illuminate our point:

Another way to unsettle the fundaments of the A-IC, especially with the younger ages, is facilitating for the animals to be recognized. In our experience, this can come in more naturally with young children than among older students who might dismiss this as anthropomorphic. For this purpose, we consider the concept of agency in Jason Hribal's texts (2007, 2011) particularly helpful: accounts of animals as agents capable of resisting human oppression, rebels in their own rights, rather than as solely victims, may empower animals and children alike, uniting them in their "underdog" position vis-à-vis adult society and the struggle against oppressive systems (cf. Corman & Vandrovcová, 2014). In Steiner school (where Karin has carried out part of her fieldwork) there is a very strong narrative tradition that bears witness of making children want to listen. Narratives told from the animal's perspective open up the possibility to identify and empathize with animals. Following a lesson like this a 10-year old student reflected: "It is kind of unfair: we pet our cats but hunt the deer. Like they were predestinated for this." This kind of student response—showing critical understanding of animals as socially constructed for human self-interest—indicates that dialogue and reflection are possible among young students, also concerning animals in the production system, and that it is mainly a matter of finding space for such dialogue.

## *Conclusion*

As the A-IC has multiple creative strategies of reaching out to schoolchildren, students, and teachers and reproducing itself through the education system, critical animal pedagogy educators must be equally creative in working with reflective, counterhegemonic pedagogical practices that respect the interests and background of each student. Together with students, we need to find ways of challenging speaking silences in education that simultaneously work to obfuscate and perpetuate animal oppression. Above all, we must carefully attend to, and support, student resistance and inquiry, whenever it may occur.

In the present education system, we thus need to eke out multiple spaces for scrutiny, critique, and rebellion against the A-IC. In our role as teachers, activists, parents, students, and children, we need to disturb. As teachers and activists we also need to consider how to bridge the gap from where we are to where we want to be in order to situate and act on these disturbances most effectively. This includes analyzing, together with students, the multiple stories and myths surrounding the animal production system, myths that thrive not only on our passive consumption of the products of this system, but also on the speaking silences that sustain them.

Challenging these myths demands creative counterstories: vegan counterstories, which reclaim the euphemistic narratives of the origins of food and use them as tools for critical exploration, for making animals heard and alternatives visible. It is to these counterstories we now turn.

## *Vegan Education: Toward Veg-Normativity?*

With vegan education, we seek a powerful statement against the A-IC by creating new conditions for learning a different kind of eating. We emphasize the learning taking place within or parallel to the larger framework of a different kind of *reading*—that is, reading (educational) texts in order to grapple with how the idea of "the human" has led us to harmful, even oppressive relations with the world (Snaza, 2013). However, new ways of reading and scrutinizing human exceptionalism (Haraway, 2008) may be difficult to incorporate into the present education system since they rely on a critique of the very foundation of this system. New ways of eating, on the other hand, may work their way into formal education in modes that appear less threatening. They may appeal to wishes to eat in a healthier and more sustainable way, catering to the sensory, palate, and aesthetics (Cole, 2014; Reclus, 1901). Still, however, many schools—and we are tempted to say that this is an inherent feature of the very nature of schooling—will resist vegan education on the grounds that it is politically controversial, coercive, and threatening to the hegemonic meat culture in society (cf. Potts [Ed.], 2016) and, above all, in most students' homes.

The key pedagogical problem we address in this section is, how can (ethical) veganism at all be taught, if the schools do not actively ask for it? Is veganism perceived as too risky to introduce to students whose parents are likely to be meat consumers?

## *Meat-normativity versus Veg-normativity*

An example of vegan education is the "Vegonorm" (Veg-norm) campaign that started in Sweden in 2016. The campaign challenges "meat normativity," referring to the institutions, structures, relations, and acts upholding the norm of production and consumption of animals, especially as everyday "meat" (Gålmark, 2005). The "Vegonorm" campaign aims to establish veganism or vegetarianism as a standard and is directed at the public but more specifically aimed at those who provide public meals such as municipalities, schools, hospitals, and homes for the elderly. This is how "veg-norm" is interpreted in relation to education in the editorial of a major Swedish newspaper:

> Today a diet based on animal foods is normative. A vegan diet an anomaly. If we are to combat global warming and unhealth (and additionally not cause animals pain) we need to start regarding vegan food as the base and food from animal sources as an (superfluous) extra. A vegan diet can supply us with all the nutrients that the body needs. We need to turn the reasoning around. Why should schools need to promptly comply with those that want to add animal products to the vegan base diet? (Nyström, 2016, our translation)

We find the idea and concept of "veg-norm" pedagogically interesting and will let it guide our discussion, although the subsequent examples will show that Swedish schools are very far from adopting veganism. Furthermore, making space for veganism, especially ethical veganism, from within daily school practice, might be a challenge. The teachers we interviewed in our fieldwork who identified as ethical vegetarians (we came across no vegans) admitted that they did not share their reason for being veggie with their students, as they felt they should not influence their students' own choices. To avoid questions, one teacher whom Karin interviewed actually said to her students that she was allergic to meat. A teacher Helena interviewed claimed that she usually waits for her students to develop their own standpoints first, and then toward the end of the class she might reveal her own (vegetarian) position. In another context, feedback from Meat Free Mondays' public speaker (J. Paulsson, personal communication, March 17, 2014) also confirmed this anxiety: when invited into school he is actively asked not to talk about the situation of animals but only environmental and health issues. In a meat-normative society, introducing veganism will likely be viewed as coercive: "I think everybody should be allowed to eat whatever they want," thus celebrating a quasi-"free choice," is an example of a student comment in the teacher education classes Helena teaches. This kind of response calls out for in-depth exploration and engagement with how certain norms become established as such and what ideological work they do.

> Lunch was a productive pedagogical exercise that day: I learned about thought experiments (although I didn't know the term thought experiment until many years later). I learned about subtext (although I didn't know the term subtext until many years later). I learned again that although school can be an expansive place where students are exposed to both new ideas and stark truths about injustices within the established order of our society, at the same moment those same injustices are systematically normalized for the sake of convenience: Mrs. Butcher would have had an easier time at lunch that day if I had just eaten the dead cow like everyone else. (Truman 2016, 39)

Sarah Truman's personal accounts of experiences from kindergarten (as quoted earlier) and primary school (above) offer examples of how meat normativity may operate in a school setting, but also indicate how the education system itself incidentally produces resistance against such norms: mainstream education may thus, potentially, become "vegan education" simply by working according to its own standards. *What is taught is not necessarily what is learnt.* Veganism has long been, more than anything, a form of rebellion and resistance against domination and therefore against the school as such. Karin's transition to becoming vegetarian and later vegan was definitely part of this tradition and in her work with children and teenagers she comes across it over and over again. While being the columnist of veggie food in an animal rights magazine and then editor of the same AR group's children's magazine she received many letters from children wanting to turn veggie against their parents' (and schools') will. In the 1980s all children in Sweden had to acquire consent from not only their parents but also the school nurse or doctor in order to be allowed vegetarian food in school. Today ovolactovegetarian food is served on demand but some schools still refuse to serve vegan meals. Work toward veg-normativity certainly supports those children and facilitates their transition to an animal-free diet.

> In an interview a 10-year-old girl in the Steiner school recounts that her father says she cannot be a vegetarian, that he says there is not enough nourishment in vegetarian food, this despite the fact that she eats vegetarian food every day at school. (Gunnarsson Dinker, forthcoming)

When it comes to veg-normativity in education, the relationship between theory and practice is complex and far from straightforward. In our experience, introducing veganism in schools is not enough. Steiner schools, for instance, are often chosen by vegetarians and vegans because they offer more vegetarian options than the average school, but this does not automatically create a deeper understanding of the situation of animals in human society, the animal production system, or human–animal power arrangements. Vegan issues are not necessarily more established in Steiner circles than anywhere else (at least not with parents and teachers). Humanist values in Steiner schools may indeed be even firmer than in state schools and the possibility of new readings of our role in the universe more limited (Gunnarsson Dinker, forthcoming). Thus, new ways of reading (Snaza, 2013) *and* eating need to exist simultaneously, and cross-pollinate each other in critical animal pedagogy.

### *Conclusion*

Previous research in critical animal pedagogy emphasizes the importance of sharing vegan food in class (Andrzejewski, 2003). While undoubtedly valuable in a pedagogical context fully devoted to animal advocacy, in our experience, moves toward veg-normativity in conventional education are a complex process. It is clearly not enough to introduce vegan foods in school settings, for instance, by token initiatives such as "Meat-free Mondays." Self-censorship employed by vegetarian teachers, or resistance against exploring the nature and work of norms in general in the classroom (cf. Bromseth & Darj [Eds.], 2010), may complicate the relation between theory and practice in critical animal pedagogy. Schools with a veggie profile, such as Steiner schools, may be even more firmly entrenched in humanist values that prevent students and teachers from critically analyzing ideas of human exceptionalism and the role of animals in human society. On the other hand, we recognize that teaching and learning is not a controllable or predictable process. Just as vegan education may produce initial resistance and possibly even (at least in the short term) sustain the meat norm, "mainstream" education may similarly give rise to critical responses among students in favor of veganism or vegetarianism (cf. Pedersen, 2010). It is in this zone of unpredictability of education that critical animal pedagogy needs to find a place to work.

## ***Critical Animal Pedagogy and the Anthropocene: Concluding Remarks***

> As educators, activists, and anarchists, we feel it our duty to cultivate vigilant critique of social norms and challenge conventional pedagogical models while also destabilizing the authoritarianism of the traditional schoolhouse setting. Following the principles of Critical Animal Pedagogy (CAP), we wish to draw attention to largely ignored and oppressive cultural norms that continue for financial and traditional reasons. (Drew & Socha 2015, p. 171)

We follow Drew's and Socha's call for the significance for critical animal pedagogy educators to deconstruct oppressive social norms in education, as well as conventional pedagogical models that serve to uphold such norms. With our chapter, however, we also emphasize the necessity to analyze the multiple roles of education in the A-IC and, in extension, the biopolitical landscape as such (Pierce, 2013).

In this biopolitical landscape, not only animal and human suffering, but also problems of climate change and global environmental damage are becoming increasingly acute. This raises a new set of questions for critical animal pedagogy: How are various categories of humans and animals at the same time implicated in and affected by climate change? How can we engage with the paradox of an ongoing mass-extinction of wild animal species taking place in parallel with an increasing mass-production of domestic animals for the food industry? How has global consumption of factory-farmed chicken become a key indicator of the Anthropocene—the epoch of human domination over the planet (Carrington, 2016)? How does the idea of "human exceptionalism" (Haraway, 2008) work in relation to these issues—and what about the role of capitalism as a driving force behind human, animal, and environmental exploitation? These are all urgent questions for educational inquiry. We believe that the intersections of animal liberation, social justice, and climate change remain to be fully explored, and suggest this is necessarily collective work to be carried out *together with* students. Due to the provisional knowledge status of these issues, at least at university level we would suggest a format of student-led seminars, as a basis for collective knowledge development. This is a topic for further exploration in critical animal pedagogy teaching and research.

Can education change the world? Can critical animal pedagogies change the world? What power does education actually have to transform not only individual lives, but society as such? As critical education scholars, we find it naïve to ascribe force and impact to education as if it would be possible for education to achieve such changes on its own (cf. Moran & Kendall, 2009): New, progressive pedagogical approaches and strategies are not enough (and can surely even be counterproductive). We need, rather, new legislation, new political and economic systems—indeed, we need a new nonviolent, anti-speciesist, postanthropocentric world order. We firmly believe, however, that education can, and must, be a part of this urgent work—not in the form of yet another programmatic and predefined curriculum, but through collective, reflective pedagogical theory and practice that disturb and disrupt oppressive arrangements and jointly create new world-forming practices for animal and human liberation.

## References

Andrzejewski, J. (2003). Teaching animal rights at the university: Philosophy and practice. *Journal for Critical Animal Studies, 1*(1), 16–34.

Aronowitz, S., & Giroux, H. A. (1986). *Education under siege: The conservative, liberal and radical debate over schooling*. London: Routledge.

Biesta, G. J. J., & Säfström, C.A. (2011). A manifesto for education. *Policy Futures in Education, 9*(5), 540–547.

Bromseth, J., & Darj, F. (2010). *Normkritisk pedagogik: Makt, lärande och strategier för förändring*. Crossroads of Knowledge 14, Centrum för genusvetenskap. Uppsala: Uppsala Universitet.

Carrington, D. (2016). How the domestic chicken rose to define the Anthropocene. *The Guardian*, 31 August. Retrieved September 16, 2016 from https://www.theguardian.com/environment/2016/aug/31/domestic-chicken-anthropocene-humanity-influenced-epoch

Cole, M. (2014). '*The greatest cause on Earth*': The historical formation of veganism as an ethical practice. In N. Taylor & R. Twine (Eds.), *The rise of critical animal studies: From the margins to the centre* (pp. 203–224). London and New York, NY: Routledge.

Cole, M., & Stewart, K. (2014). *Our children and other animals: The cultural construction of human-animal relations in childhood*. Farnham: Ashgate.

Corman, L., & Vandrovcová, T. (2014). Radical humility: Toward a more holistic critical animal studies pedagogy. In A. J. Nocella II, J. Sorenson, K. Socha, & A. Matsuoka (Eds.), *Defining critical animal studies: An intersectional social justice approach for liberation* (pp. 135–157). New York, NY: Peter Lang.

Drew, L. (2014). Embodied learning processes in activism. *The Canadian Journal for the Study of Adult Education, 27*(1), 83–101.

Drew, L., & Socha, K. (2015). Anarchy for educational praxis. In A. J. Nocella II, R. J. White, & E. Cudworth (Eds.), *Anarchism and animal liberation: Essays on complementary elements of total liberation* (pp. 163–178). Jefferson, NC: McFarland.

Gålmark, L. (2005). *Skönheter och odjur: En feministisk kritik av djur-människa-relationen*. Göteborg och Stockholm: Makadam.

Giroux, H. (1981). *Ideology, culture and the process of schooling*. Philadelphia, PA: Temple University Press.

Gunnarsson Dinker, K. (forthcoming). *Exploring the concepts and practicalities of "use" and "respect" in relation to animals: Case studies in Swedish primary education*. Doctoral thesis manuscript in preparation. Swansea: Swansea University.

Gunnarsson Dinker, K., & Pedersen, H. (2016). Critical animal pedagogies: Re-learning our relations with animal others. In H. Lees & N. Noddings (Eds.), *Palgrave international handbook of alternative education* (pp. 415–430). London: Palgrave Macmillan.

Haraway, D. J. (2008). *When species meet*. Minneapolis, MN and London: University of Minnesota Press.

hooks, B. (1981). *Ain't I a woman: Black women and feminism.* Boston, MA: South End Press.

Hribal, J. (2007). Animals, agency, and class: Writing the history of animals from below. *Human Ecology Review, 14*(1), 101–112.

Hribal, J. (2011). *Fear of the animal planet: The hidden history of animal resistance.* Oakland, CA: AK Press.

Hussar, K. M., & Harris, P. L. (2009). Children who choose not to eat meat: A study of early moral decision-making. *Social Development, 19*(3), 627–641.

Illich, I. (1970). *Deschooling society.* New York, NY: Harper and Row.

Kopnina, H., & Cherniak, B. (2015). Cultivating a value for non-human interests through the convergence of animal welfare, animal rights, and deep ecology in environmental education. *Education Sciences, 5*, 363–379.

Linné, T., & Pedersen, H. (2014). 'Expanding my universe': Critical animal studies education as theory, politics, and practice. In J. Sorenson (Ed.), *Critical animal studies: Thinking the unthinkable* (pp. 268–283). Toronto: Canadian Scholars' Press.

Linné, T., & Pedersen, H. (2016). With care for cows and a love for milk: Affect and performance in dairy industry communication strategies. In A. Potts (Ed.), *Meat culture* (pp. 109–128). Leiden: Brill.

MacCormack, P. (2013). Gracious pedagogy. *Journal of Curriculum and Pedagogy, 10*(1), 13–17.

Malm, K. (2007). *REDE: Respekt, Empati, Djur, Etik. Ett värdegrundsmaterial med målet att utveckla barns empatiska förmåga och respektfulla synsätt gentemot djur och natur.* Djurskyddet Sverige och Svenska Djurskyddsföreningen. Retrieved from http://www.rede.se/

McLaren, P., & Kincheloe, J. (2007). *Critical pedagogy: Where are we now?* New York, NY: Peter Lang.

Moran, P., & Kendall, A. (2009). Baudrillard and the end of education. *International Journal of Research & Method in Education, 32*(3), 327–335.

Myers, O. E., Jr. (2013). Children, animals, and social neuroscience: Empathy, conservation education, and activism. In M. Bekoff (Ed.), *Ignoring nature no more: The case for compassionate conservation* (pp. 271–285). Chicago, IL: The University of Chicago Press.

Noske, B. (1997). *Beyond boundaries: Humans and animals.* Montréal: Black Rose Books.

Nowatzki, A. (2013). Vegan parenting: Navigating and negating speciesist media. In K. Socha & S. Blum (Eds.), *Confronting animal exploitation: Grassroots essays on liberation and veganism* (pp. 89–111). Jefferson, NC: McFarland.

Nyström, D. (2016, February 10). Kött är en het potatis. *Gefle Dagblad*, Retrieved from http://www.gd.se/opinion/ledare/kott-ar-en-het-potatis

Palmer, A., & Pedersen, H. (2017). Naturhistoriska riksmuseet som kritisk djurpedagogisk arena. In B. Bergstedt (Ed.), *Posthumanistisk pedagogik: Teori, undervisning och forskningspraktik* (pp. 41–57). Malmö: Gleerups.

Pedersen, H. (2010). *Animals in schools: Processes and strategies in human-animal education.* West Lafayette, IN: Purdue University Press.

Pedersen, H. (2011a). Animals and education research: Enclosures and openings. In P. Segerdahl (Ed.), *Undisciplined animals: Invitations to animal studies* (pp. 11–26). Newcastle upon Tyne: Cambridge Scholars Publishing.

Pedersen, H. (2011b). Counting affects: Mo(ve)ments of intensity in critical avian education. *Canadian Journal of Environmental Education, 16*, 14–28.

Pedersen, H. (2012). Undercover education: Mice, mimesis, and parasites in the teaching machine. *Studies in Philosophy and Education, 31*(4), 365–386.

Pedersen, H. (2013). Follow the Judas sheep: Materializing post-qualitative methodology in zooethnographic space. *International Journal of Qualitative Studies in Education, 26*(6), 717–731.

Pedersen, H. (2015). Parasitic pedagogies and materialities of affect in veterinary education. *Emotion, Space and Society, 14*, 50–56.

Pierce, C. (2013). *Education in the Age of biocapitalism: Optimizing educational life for a flat world.* New York, NY: Palgrave Macmillan.

Potts, A. (Ed.). (2016). *Meat culture.* Leiden: Brill.

Reclus, E. (1901). On vegetarianism. Retrieved from myweb.tiscali.co.uk via Anarchist Library.

Rowe, B. D. (2011). Understanding animals-becoming-meat: Embracing a disturbing education. *Critical Education, 2*(7). Retrieved from http://m1.cust.educ.ubc.ca/journal/index.php/criticaled/article/view/132

Rowe, B. D. (2012). *Consuming animals as an educational act.* Unpublished doctoral thesis. Ohio University EDU Policy and Leadership.

Rowe, B. D. (2013). It IS about Chicken: Chick-fil-A, posthumanist intersectionality, and gastro-aesthetic pedagogy. *Journal of Thought, 48*(2), 89–111.

Snaza, N. (2013). Bewildering education. *Journal of Curriculum and Pedagogy, 10*(1), 38–54.

Sobel, D. (1996). *Beyond ecophobia: Reclaiming the heart in nature education.* Great Barrington, MA: Orion Society.

Stewart, K., & Cole, M. (2009). The conceptual separation of food and animals in childhood. *Food, Culture and Society, 12*(4), 457–476.

Svärd, P. A. (2015). *Problem animals: A critical genealogy of animal cruelty and animal welfare in swedish politics 1844–1944.* Unpublished doctoral dissertation. Stockholm Studies in Politics 165. Stockholm: Stockholm University.

Taje, E. (2016). Destroy the school and create a free school digging up the roots of dominant/submissive complexes and planting the seeds of cooperative social interaction. In S. Springer, M. Lopes de Souza, & R. J. White (Eds.), *The radicalization of pedagogy: Anarchism, geography, and the spirit of revolt* (Kindle Locations 4266–4782). Rowman & Littlefield International. Kindle Edition.

Tiqqun. (2011). *This is not a program.* Los Angeles, CA: Semiotext(e).

Trachsel, M. (2017). The presence of "pork" and the absence of pigs: Changing stories of pigs and people in Iowa. In D. Nibert (Ed.), *Animal oppression and capitalism* (Vol. 1, pp. 110–140). Praeger.

Truman, S. E. (2016). School sucks for non-human animals. *Journal of Curriculum and Pedagogy, 13*(1), 38–40.

Twine, R. (2012). Revealing the 'animal-industrial complex'—A concept & method for critical animal studies? *Journal for Critical Animal Studies, 10*(1), 12–39.

Yates, R. (2005). *The social construction of human beings and other animals in human-non-human relations.* Unpublished doctoral dissertation. University of Wales, Bangor, UK.

Yates, R. (2013). *Growing up as animal harming animal lovers: Sociology and animal use* [video presentation]. Retrieved from http://veganinformationproject.org

# 4. *Our Heroes Need to Wear Ski-Masks: The Animal Man and the Animal Liberationist Hero in Comics*

SEAN PARSON

Superhero narratives have become central to how our culture understands heroism. We think of caped and masked men fighting street crime, working to thwart a sinister plot for world domination, or stopping an alien invasion. We picture Batman, or Superman, or the X-Men. Many like to claim that these stories—especially considering how fantastical the stories and powers are—are mere escapism, that they primarily serve as a way to find comfort in an otherwise chaotic and scary world, but that is far from the truth. Superhero stories, as Eco and Chilton (1972) note in Superman, can serve as a tool to protect the status quo and protect the rule of property, but they can, as Ramzi Fawaz (2016) notes, also open up space for creative imagination, alternative worlds, and expanded democratic involvement.

It is partially through the stories we see in comic books and superhero films that our culture develops and generates its understanding of what it means to be a hero. Throughout much of its history the genre of comics can be seen as creating a space for both reactionary and radical politics to emerge and the heroes that emerge from this process tend to embody these politics: we see both Iron Man, the defender of the American Military Industrial Complex (Costello, 2009), and Captain America, the embodiment of American liberalism (Dittmer, 2013), but we also see Magneto, the troubled defender of a group facing genocide, and early Superman, whose tagline was "champion of the oppressed" (Saunders, 2011). While many comic book series foreclose certain conversations around animal abuses and systemic violence in our food, medical, and entertainment industries, and in doing so naturalized and justified—often through its erasure—carnism and speciesism, there is a long history of comics justifying and defending speciesism; there are a strand of comics—starting with Grant Morrison's work on Animal Man—that has expanded the superhero discussion to include nonhuman animals. It is these counternarratives that provide a pedagogical space to explore what it means to be a hero for nonhumans and to create a broader discussion around animal liberation issues in general. In this chapter, I am going to look at Grant Morrison's iconic animal liberation story, Animal Man, by putting his 26-issue arc in dialogue with work in critical animal studies and total liberation. In doing this, I argue that comics can serve as a form of educational space outside the classroom and therefore are a means of engaging with critical ideas.

## *Pedagogy Outside the Classroom*

While it is often assumed that teaching only occurs within the classroom, such a generalized statement is not only shortsighted but also wrong. As Richard Kahn (2010) argues we need to expand our understanding of pedagogy, especially in relation to the growing ecological and political crises we are facing, to include not just formal educational locations but also critical spaces for education that are created by activist. By focusing his last chapter of the book on Judi Bari, the famed Earth First! activist that linked environmental and labor issues in the 1990s, Kahn opens up what we can think of as educators to include activists as well as teachers and mentors. In thinking of education in this broader lens, we can begin to think of the many ways in our daily lives in which we are being taught a lesson, either knowingly or unintentionally.

While critical educators have historically focused on the "banking" style education model that developed during the 19th and 20th centuries, understanding this approach as a form of propaganda to maintain the capitalist order, few have expanded their educational analysis to the world of popular media. Expanding on Henry Giroux's work to link cultural studies with the concept of "public pedagogy" we can start to understand that there is a direct link between media and education. While many of those on the left understand the propagandistic role that media in general serves in this culture—constructing and reinforcing the hegemonic liberal, racial, and gendered orders—most of those on the left do not view these spaces as "educational spaces." The few, like bell hooks in *We Really Cool: Black Men and Masculinity* (2004), tend to see popular culture as essential in promoting the dominant, patriarchal, conceptions of masculinity that dominate our society. Similarly, the articles in *Hollywood's Exploited: Public Pedagogy, Corporate Movies, and Culture Crisis* (2010) also primarily focus on the problematic and negative impact of popular culture—focusing on representations of disability, capitalist values, and racial caricatures. Even Giroux's book *Zombie Politics and Culture in the Age of Casino Capitalism* (2015), when it discusses and engages public pedagogy, mostly sees it as a form of elite propaganda that helps zombify the American public, allowing for the breakdown of democratic social and political spaces.

Critical pedagogues such as Anthony J. Nocella II, Lara Drew, Richard Kahn, Lauren Corman, and Constance Russell have made the connection between critical pedagogy and animal liberation. The field of critical animal studies seeks to provide a radical and interdisciplinary analysis and critique that links systems of oppression while at the same time providing a moral voice defending radical activism—like the Animal Liberation Front. Anthony Nocella argues that one of the roles of education is not only to both unmask the powerful and make their violence appear clear, but also to learn about, and come to understand, the reasons that many activists turn to acts of property damage and liberation (Nocella, 2007). This can be accomplished by creating pedagogy that not only critiques dominant culture and politics but also provides an in-depth analysis of radical organizations and movements.

In this piece I am not arguing that this critical perspective is wrong, as popular culture certainly does serve as a tool of propagandistic control; instead, I claim that this vision of popular culture is incomplete. In addition to serving as a tool of control, popular culture also can, and regularly does, serve as a critical response to the dominant perspectives in our society. In this regard, some aspects of popular culture can and should be understood as forms of critical education, providing a lens and perspective that undermines the dominant species, class, race, and gender narratives. One such example of this is the superhero comic *Animal Man*, but before we get to that we need to explore the complex relationship between critical and status quo politics within the American comic book industry.

## *Comics as Liberatory Pedagogy and as Reactionary Propaganda*

In the early era of comics, the genre had nearly universal readership, with an estimated 90% of youth, including girls, reading at least one comic on a weekly basis. This was not as big of a concern for the elites in society during World War II, as the comic industry become primary defenders of the American war plan—with every major character, from Superman and Wonder Woman to Batman and Captain America, involved in the war effort in some part. The role these comics played during World War II has, recently, come under intense scrutiny, as the series tended to embrace racialized caricatures of the Japanese, turning Japanese characters into rat-like monsters (Wright, 2001). Likewise, during the Cold War the comic book industry both embraced and defended the Cold War liberal consensus. Characters like Iron Man originally fought for the U.S. against the Viet Cong and Captain America villains, like the Red Skull, where transformed form Nazi's to communists (Castello, 2009).

While it might be easy to assume that since the comics followed the norms of the dominant society, they primarily served as forms of propaganda, shaping the minds of the youth, this analysis would be overly simplistic. As Ramzi Fawaz (2016) showed in his masterful book *The New Mutants* the superhero comic world also provided an educational space that allowed for the imagining of utopian possibilities and the expansion of democratic norms and values. In this work Fawaz looked to a range of comics, but it was his engagement with the 1980s *X-men* comics and *Fantastic Four* that is most valuable for this discussion. In his analysis of *X-Men* he argues that the characters stand in for the general outcasts of society—sexual, racial, religious, and others—and provides a space for them to begin thinking of their difference in terms of uniqueness, while the *Fantastic Four* highlights the queer beauty of being different. Whether it is the Amazonian black beauty of the X-men Storm or the flaming queer Human Torch, Fawaz shows us that comics provide a space to question dominant norms around race, gender, sexuality, and ability.

Beyond just showing us representations of difference, comics are also a form of education. The founder of Wonder Woman, the eccentric William Marston, understood that the comic form could have a profound educational effect on children. What he wanted to do with Wonder Woman was teach the youth about feminism, by showing them a powerful female figure, which solved problems with cunning and communication more often than with her fists (or Lasso) (Berlatsky, 2015). It was the educational role of comics, and their ability to transgress social boundaries, that led to the psychologist Fredrick Wertham (1954) to write the bestselling book *Seduction of the Innocents.* This work argued that the comic industry promoted and taught children asocial behavior—from the acceptance of fascism and vigilante violence, to the normalizing of homosexual behavior and sadomasochism. Some of these best examples of the danger of comics came from Wonder Woman, who he correctly saw as promoting a sado-masochistic (S & M) sexual desire (one that the founder admitted to), and from Batman, where Wertham argued that the Batman and Robin relationship was a wish fulfillment for gay men. In both cases, Wertham saw comics as teaching an alternative form of sexual relationship and he feared that by educating the youth that S & M or homosexuality is acceptable, it would prompt more youth to experiment. While Wertham provided nothing but anecdotal evidence that the violent crime and superhero comics corrupted the youth he did inspire a conservative social movement around the nation that worked to ban or censor comics. The backlash led to the comics' code—an industry-imposed censorship board—that for decades censored overt sexual, racial, or transgressive discussions in comics. Instead of overt discussions of these topics the comic industry found more creative ways of discussing these topics, using aliens, superheroes, and insane schemes to explore issues that were central to popular culture. This means that for most of the superhero genre's history there has been a tension between transgressing social norms and reinforcing social and political consensus. This back and forth dialect makes comics, as well as other forms of popular culture, important sites of exploration for academics, activists, and educators.

In effect, what I am arguing here is that comics, much like most of popular culture, cannot be understood simply as a form of propaganda. It is true that dominant media tends to reinforce the cultural values and norms of society but if that is all it did the end result would be stale. In order to create a sense of excitement in a story the characters have to engage with topics and issues that people connect with: people's fears, anxieties, and desires. It is this engagement that, in a well-done comic, an educational space can be created to explore difference and expand democratic spaces. It is this dialectic between the status quo and the other that allows for a space of engagement and growth within the genre. That said, I do not want to romanticize the liberatory potential of popular culture, and comics in general; in most instances even the stories that transgress and expand the moral world view often still reinforce dominant cultural values and norms.

For the rest of this chapter I want to explore an example of how comics can both open up a space for engagement and discussion and also reinforce cultural values at the same time by exploring the ways in which comics have attempted to expand the notion of heroism beyond the human to include animal liberation as a heroic act. In doing this, I will critically explore the iconic series *Animal Man*, by Grant Morrison, and bring this work into dialogue with the work in critical animal studies and the position of total liberation.

## *Background on Animal Man*

Buddy Baker, while on a hunting trip with his friends in the 1960s, comes across a downed UFO in the backcountry forests. While investigating the crash, the UFO explodes, seemingly killing the young man. Miraculously he does not die but instead the accident gives Buddy super powers, which allow him to "borrow" the powers of animals that surround him. Meaning he could see a bird and gain its ability to fly (although he does not sprout wings) or he could see a fish and borrow its ability to breathe underwater (without growing gills). The explosion turned Buddy Baker into Animal Man, "the man with animal powers!"

The character was originally created by David Wood for the comic *Strange Adventures* #180 (1965) where the character was a surprise hit and was regularly featured over the next two years. After his run in *Strange Adventures* Animal Man would sporadically appear in comics throughout the 1970s and 1980s, once battling Wonder Woman and another time helping Superman. He was even a member of the short-lived "super group" of minor characters called "The Forgotten Heroes."

The stories and comics that Animal Man starred in during the first three decades of his existence have been, for the most part, completely forgotten, and it was not until Grant Morrison, a critically acclaimed British author who had just had a successful run on Batman, convinced DC to let him take the character and develop a monthly series about his adventures. Partially due to the success of Neil Gaimen's *Sandman*, which also resurrected a forgotten character from DC's past catalog, DC comics agreed. The series was originally created as a vehicle for Morrison to work through his own ideas and concerns regarding animal rights, as he had just recently become a vegan animal rights activist, and it also allowed him (and DC) to push the boundaries of what a "superhero" is.

In Grant Morrison's hand the character was given a deeper backstory and his origins and powers are further developed. Most importantly, Morrison changed Animal Man's powers, as he no longer "borrowed" from animals around him but he instead connected into the "morphogenic field," which is never fully described but seems to be something like a platonic transcendental space where the "forms" of all animals exists. While this "morphogenic" field sounds absurd, it allowed Morrison the ability to discuss the idea of transcendental reality and also allowed him to explore biocentrism. It is the first of these aspects—the ability to discuss transcendental reality—that has garnered the majority of academic discussion on Animal Man. Grant Morrison's run is mostly known for its unique narrative style, with its intentional breaking of the fourth wall, the introduction of Grant Morrison himself as a character in the story, and the overall discussion around issues of free will, authorship, and continuity. What has been lost in academic readings of the series is the way it engaged with animal rights.

## *Themes and Concepts in Animal Man*

In this section I want to put the speculative philosophic nature of Grant Morrison's Animal Man in dialogue with the work in Critical Animal Studies. This section will focus on three broad themes found in Animal Man that relate to animal liberation topics. They are: The role of empathy and care in animal activism, the interconnected nature of all life, and the role that power plays in constructing anthropocentric institutions. What should be clear by the end of this conversation is that the pages of Animal Man serve as a critical classroom for engaging with ideas and concepts central to the animal liberation movement.

## *The Power and Limits of Empathy as a Political Tool*

Anyone who has ever seen an animal rights documentary or been given a "why vegan" flier will know that empathy and compassion play central roles in the animal rights movement. This emphasis is not new, as the 19th-century humane societies and anti-vivisection groups highlighted the suffering of animals in their work. Many radical theorists of the time also embraced animal rights views, such as Peter Kropotkin and Leo Tolstoy, but it was the writing of the anarchist geographer Recluse who developed some of the earliest theories around political vegetarianism. To Reclus, writing in 1902, the treatment of nonhumans by our society is barbaric, and separates us from the intimate and respectful connection we should have with nonhumans. As Reclus writes,

> …for the great majority of vegetarians, the question is not whether their biceps and triceps are more solid than those of the flesh-eaters, nor whether their organism is better able to resist the risks of life and the chances of death, which is even more important : for them the important point is the recognition of the bond of affection and goodwill that links man to the so-called lower animals, and the extension to these our brothers of the sentiment which has already put a stop to cannibalism among men. (Reclus, 1901)

Here Reclus, more than 60 years prior to the publication of *Animal Liberation* and the rise of the modern Animal Rights movement, makes an emotional appeal for vegetarianism based on an empathic concern for animal welfare and ties that empathy to a "higher state" of thinking.

Peter Singer in his seminal work *Animal Liberation* (1975) also placed empathy at the center of his animal rights ethics. Peter Singer argues that utilitarianism, which argues that the ethical right choice is that which causes the greatest benefit for the greatest number, when deciding who to include, its ethical calculation should not only focus on humans but instead all sentient beings. He claims that we should not exclude a being because they are not "rational" or that they lack language. Instead we should not ask can they talk, or think, but instead we should ask, "Can they suffer?" To Singer suffering is sufficient for a being to deserve ethical consideration and that we should ethically act in a way that minimizes suffering. While Singer here recognizes the importance of empathy he removes the emotional appeal to it by formalizing the empathy within a cold, calculating, ethical formula. However, abstracting the suffering of nonhumans and creating rational formulas for ethics has been criticized by feminist animal rights theorist and activists, such as Kim Socha (2012), Carol Adams (1990), and Pattrice Jones (2005). As Socha writes, "animals don't like to be tortured, and that is reason enough not to torture them. How do we really know? We know because we just know. We don't have to prove it through the language of patriarchy" (Socha, 2012, p. 233). These feminist critics argue that Singer undervalues the importance of emotion and empathy in animal rights politics. Instead of looking to abstract theories and a rational defense, they argue that an empathetic ethic of care is all that is needed to justify an animal rights politics.

The discussion and debate around empathy is also worked out in the pages of Animal Man where the character highlights both the strengths and limitations of empathy in constructing an animal rights politics. In Animal Man Buddy Baker's empathy is tied to his powers because he not only borrows powers from nonhumans but also shares in their experiences. In the second issue of the series (1988) the character is recruited by a research lab to investigate a break-in, and while in the lab Animal Man develops a connection to the monkeys and dogs in the research lab. In borrowing a dog's excellent sense of smell, he laments the poor health of the dog, and while looking at a group of monkeys fused together into a monstrous chimera he experiences the pain and suffering and fear they have experienced while being prisoners in the lab. This empathy is further strengthened as the character increasingly comes to understand that the supposed "villain" who broke into the lab, the B'Wana Beast, is in fact a hero trying to stop the lab from experimenting on his friend—a human/ape hybrid.

By the end of the first story arc (issue #5) Animal Man is convinced that the medical lab, which is attempting to develop a strain of anthrax for the military that kills only livestock and not humans—is the real villain in this story. In learning and connecting with the B'Wana Beast, who is infected by anthrax after his human/ape friend dies from the disease, Animal Man saves him from the illness and confronts the lab scientists, referring to them as monsters. After Animal Man leaves the lab, the B'Wana Beast (who has the power to "fuse" animals together) fuses the head researcher with his dead human/ape friend so the man can experience vivisection as penance for his crimes.

The experiences Animal Man has with lab animals (which is a common theme) makes him realize that the suffering of animals in our society is so great that he needs to do something to stop the violence. The horrors and brutality of medical experimentation are not glossed over in Animal Man and the suffering of the animals and the outrage and anger of Animal Man force the reader to think about the lived experiences of lab animals, and to also question their own complacency in allowing the suffering to occur. Within the narrative, it is the experience of lab animals that prompts a larger change in Buddy Baker's life as he comes to realize that the suffering of monkeys and dogs in labs is similar to, if not worse than, the suffering of animals slaughtered for food or hunted as trophies. The first thing he does is decide to go vegan—throwing out all the animal products in his family's home to the anger of his wife Ellen and son Cliff, who are not convinced by his ethical arguments. He also decided to become a fulltime superhero for animal rights groups—working with the equivalent of the Sea Shepherds and ALF and also working with hunt saboteurs to stop a foxhunt in England.

Quickly his radical activism in defense of animals—which is based on empathy with their suffering—puts him in conflict with his friends, his family, and even his manager. Upon becoming vegan Buddy Baker has a fight with his wife who finds his opposition to his family eating animal products oppressive to her ability to "choose" what to eat and also has multiple arguments with his son over veganism as his son is always sneaking off to eat animal products. Animal Man's activism with the ALF and other radical animal rights groups is too much for his manager who sees his activism, most notably a raid to release animals from a medical lab, as morally indefensible, controversial, and bad business. In response to the attacks brought forth by his manager Animal Man asserts that "You're just assuming that a rat's life is somehow less important a human life. Who's to say that is true? Think about it for a minute what makes a human being any more important to the universe than a rat?" Here Animal Man confronts his manager's anthropocentrism by highlighting the rational argument that nonhumans have intrinsic value and should not be tortured or experimented on. Much like with his family, the argument fails to convince and the manager quits. In these cases, Buddy Baker is not able to convince anyone in his family except his youngest daughter Maxine, who is just as empathetic as he is and later on was shown to have even more powerful Animal powers. His wife and son as well as his manager and neighbors are unable to experience the suffering of nonhumans to the same degree as Buddy and as such are blind to the problems that nonhumans experience.

Lastly, Animal Man does not describe empathy as a human emotion. In fact, it regularly highlights how distorted the human sense of compassion and empathy is, often by highlighting how empathetic most nonhumans naturally are. In this regard Animal Man seems to echo the work of Rousseau (2011), who in *The Origin of Inequality* argues that compassion is a natural emotion that civilization impedes. By highlighting compassion and empathy found in other species Animal Man provides a powerful critique of modern human callousness. For instance, in issue #15 (1989) Animal Man teams up with an environmental group to stop a yearly dolphin slaughter in the Faroe Islands. While attempting to stop the massacre, Animal Man fails to protect one dolphin from a native fisherman. Angered by the murder, Animal Man takes the fisherman and leaves him in the middle of the ocean to die. Luckily for him the murdered dolphin's parent swims in and saves him. While saving the native hunter, the dolphin asserts (through narration) that:

> One day only the world will exist, no more agony, no more fear in all the vast unfolding of time and the world/ One day/ Until then the killing will continue/ the blood spilling slaughter of innocents/ Until then there will be aggression and pain and sadness/that is the way/that is the way of the sad hu-men/our way is different. (Morrison, Animal Man issue #15)

The empathy of the dolphin to show care and love for this killer, the man who killed his child, is meant to highlight the emotional differences between humans and nonhuman animals. The dolphin, after experiencing such pain and suffering at the hand of a human, is not angry and resentful, but he is sad, sad for humanity.

Overall, Grant Morrison's run of Animal Man centralizes compassion and empathy and makes it the primary motivating force for the character's heroism. While Animal Man regularly interacts with other empathic characters—from the B'Wana Beast to his daughter Maxine—his wife, son, friends, and other superheroes rarely understand his commitment to stop animal suffering. In fact, other than the activists he works with, it is only the heroes who have the ability to communicate with nonhumans that share his commitment to animal liberation. In effect, the inability, or unwillingness, of humans to deeply connect with nonhuman animals is shown to be the primary barrier for compassionate activism. The rational arguments Buddy Baker uses rarely convince people that his cause is just—as is the case when he attempts to defend liberating lab animals in a TV debate with a medical researcher—instead what ends up winning over his family, and other heroes like the Martian Manhunter, is the passion and commitment the character has to stopping animal suffering; it is his passion for justice not rationality or appeals to guilt that convinces people of the rightness of his cause.

## *Total Liberation and the Interconnected Nature of Oppression*

Animal Man gets his animal powers by tapping into the "morphogenic field," which, while poorly defined in the comic, appears to connect all living beings (past and present). The field is described as both a web of life and a spiritual, transcendental world that only a few people have access to. The Morphogenic field is used to justify a total Liberation framework to animal activism. Total liberation argues that all forms of oppression are linked together and as such animal rights should be tied to human rights, which should be tied to struggles against white supremacy, colonialism, and patriarchy. The concepts of total liberation "requires us to understand the connections between human, nonhuman, and ecological liberation. It posits that human liberation is incomplete—as it would still be rooted in domination and oppression—if it does not include these other facets" (Collings, Parson, & Arrigoni, 2014).

In Animal Man the clearest example of how the morphogenic field connects life, in a total liberation style, is found in issue #13 (1989) when Animal Man works to convince an antiapartheid activist (Dominic) in South Africa to take on the mantel of the B'Wana Beast and protect African wildlife. In this issue, Animal Man and the current B'Wana Beast break into a South African prison where a white South African military leader is torturing Dominic. At first Dominic is opposed to taking on the mantel of the B'Wana Beast stating that he was uncomfortable taking on "that white imperialist title" and is uncomfortable with having the power to "control" animals. He also argues that in the past the B'Wana Beast has not fought to protect all of Africa, but only to protect nonhuman life. As an example, he tells the story of how he and his friends were tortured by the South African government simply because they wanted to be treated with respect, yet the B'Wana Beast, the supposed protector of Africa, has never done anything to stop the Apartheid government from committing these atrocities.

Dominic does decide to take the mantel, but in doing so declares that he must change his name away from the colonial and imperialist B'Wana Beast, which he associates with white oppression, and renames himself the "Freedom Beast." At this moment Dominic flips the B'Wana Beast narrative on its head and declares that as Freedom Beast he will actually fight to protect all animals in Africa and not just nonhuman wildlife. His struggle for human animals starts with he and Animal Man working together to free political prisoners in South Africa.

This narrative provides a strong critique of liberal white animal rights activism, which can be seen as emulating the politics of the B'Wana Beast (the white imperialist). That character spent his time fighting for the animals of Africa—the lions, elephants, and nonhuman primates—but never saw humans as animal needing protection and compassion. To B'Wana Beast fighting apartheid was "too political" and therefore would sully the purity of his cause. Similarly, the B'Wana Beast represents a paternalistic stewardship approach to environmental and animal rights issues. He talked about having the power to "control" nonhuman animals, which limited the character's ability to decenter humans from their position at the top of the species hierarchy.

Dominic the Freedom Beast takes a different approach. First, Freedom Beast understands the reality that humans and nonhumans are both animals and therefore as the protector of Africa's animals he has a moral obligation to stop poaching and other crimes against nonhumans, but also to stop Apartheid. In considering human oppression as part of the politics of animal liberation, Freedom Beast, well before the linkages were clearly theorized, embraced a total liberation politics. Furthermore, the Freedom Beast clearly understands the colonial and imperial relationships that have shaped African and animal rights politics. This anti-imperialist politics is reinforced, quickly, by the character's unwillingness to describe his power as being to "control animals" but instead viewing his power as being able to "talk to them." By focusing on communication instead of control, Freedom Beast alters the imperial relationship that the B'Wana Beast had with the nonhuman world.

The politics that this issue promotes is intersectional and antiauthoritarian. The comic book makes an argument that since all life is interconnected by web of relations (the morphogenic field), those who fight for animal rights invariably need to understand that humans are animals as well. In this story the struggle against imperialism, colonialism, and racism comes together and becomes intertwined with the politics of animal liberation politics which allows for the Freedom Beast to avoid the white, imperial failures of his predecessor the B'Wana Beast.

Similar aspects, connecting animal issues with human politics, exist throughout the entire Morrison run on Animal Man. For instance, in issue 9, the Justice League sends construction workers to fix damage done to Buddy Baker's house after a villain attacks him and his family at home. Throughout this issue, the workers regularly bring up issues of class (the Baker Family is always described as working class) and actually praise him for standing up for the little guy and praise him for being an animal rights hero. In one scene, an unnamed construction worker says:

> ...See your husband, he's one of us, you know? A man of the people...and it's about time one of 'em started looking out for animals, ya know? I tell you, my old man, he used to keep every sort of animal you ever heard of: Dogs, Cats, Rabbits, Snakes...house was like the San Diego Zoo. We could've charged admission.

Even though the series does not always succeed in linking human and nonhuman concerns (for instance the series reinforces native othering in its depiction of native fishers on the Faroe Islands), by attempting to explore the intersections of race, class, colonialism, and animal rights Animal Man provides an early foray into the concept of total liberation.

## *Thrasymachus Returns: Realism, Power, and Animal Exploitation*

> In the end it all boils down to three words. 'MIGHT MAKES RIGHT'. Man is able to abuse and slaughter and experiment on animals simply because he's stronger than they are. Other than that, there's no moral ground on which to justify any animal exploitation.—Grant Morrison (character), Animal Man #26 (1990)

The above quote comes from a dialogue between Buddy Baker and the "divine creator" who controls his life and story—Grant Morrison, who appears as a character in issues 25 and 26. In the final issue of Grant Morrison's run on the series, Animal Man, trying to come to grips with the death of his family, searches the afterworld to find his creator and convince him to bring back his family. The issue starts with Animal Man confronting Grant Morrison while he writes in front of his computer. The two have a long philosophic discussion around issues of free will, multiple realities, and animal rights. Grant Morrison here discusses the soapbox quality of the comic and argues, as the above quote show, that in his mind the exploitation of animals is entirely due to power; that what makes humans "superior" to nonhumans is simply "MIGHT MAKES RIGHT". He then forces Animal Man to fight an obviously absurd character he calls "The Slaughterhouse" and says "The idea is you fight this guy and you settle the moral argument by beating him into the ground. Don't laugh. That's the way we deal with things in the real world too." Here Morrison contends that power, not ideas, is what matters in protecting the animals (human and non-human) from oppression and exploitation.

While the last issue formalizes this position, this idea exists throughout the entire series. It is seen within the first story arc when Animal Man is unable to stop the research lab from experimenting on primates because they had big corporate and government contracts that protected them; their economic and political power allow them to abuse animals. Likewise, it was a corporate/government organization that hired a contract killer to murder his family in issue 20. The killer was hired because powerful men from both the US government and from corporate boardrooms feared that Animal Man would continue to undermine their economic and political interests.

The argument that might makes right should not come as a surprise for readers of superhero comics as the genre has always been criticized for arguing that violence is the easiest way to address a problem. This was one of the primary criticisms offered by Wertham in Seduction of the Innocents and has been a reoccurring theme in comic criticism ever since. In general, the genre tends to have heroes and villains symbolize values, ideas, or ideologies and the resulting fight can be seen as a battle of ideas. In the case of Animal Man, I think it would be a mistake to dismiss Morrison's analysis of power politics as being a result of genre norms because here the author is having an intentional, self-conscious, discussion about power, politics, and animal rights. So, what does it mean for animal liberation and animal rights if Morrison is correct and power, not ideas, is the central barrier to stopping exploitation? First, it means that the limits of empathy (as expressed earlier) to reach a wide range of people needs to be understood as a primary limitation on animal rights activism. If "right" and "wrong," and "justice" and "oppression," are defined by a hegemonic cultural power, what is needed is to confront those institutions of power and to create a counterhegemonic narrative. The structures of anthropocentrism are maintained not by ideas or psychological barriers but by power—our legal system allows for animals to be owned as property, economic efficiency requires them to be mistreated and abused, and the pharmaceutical industry uses their economic power (and our own fear of illness and death) to justify animal experimentation. In this regard, compassion and empathy can serve important roles and are needed aspects of the counternarrative, but first and foremost we must actively confront institutions of power that systematize animal oppression.

## *Conclusion*

As can be seen by the exploration of Grant Morrison's work on Animal Man, the superhero format can be used as a means to critical engage and explore complex animal liberationist topics. In fact, I would argue that this work serves as a form of critical animal studies pedagogy as the pages of the work served as a classroom for both those who did and did not have any direct knowledge of the animal rights movement. Each issue took aspects and debates in the animal rights community and turned them into a fun, action-packed, engaged story with complex conspiracies, overpowered villains, family drama, and an occasional fight scene. The stories that are found in Animal Man provide an ideological and philosophic discussion of animal rights, a critique of dominant speciesist culture, the development of a total liberation theory, and also insight and exploration of the passions and reasons that activists risk their lives to save nonhuman lives. Much like all superhero comics, Animal Man makes us think about what it means to be a hero. When one sees oppression and violence how does one react? Unlike the traditional superhero story, be it superman and Batman or even the X-men, our empathic lens is always directed at human (or Alien) oppression and the lived experiences of torture and suffering of nonhumans is almost always made invisible. But if you are a real hero, if you are really committed to stopping suffering and being a champion for the oppressed, Animal Man teaches us that we need to put on our black ski-masks, grab a crow bar, and free some animals from a lab or a farm.

## References


Adams, C. J. (1990). *The sexual politics of meat: A feminist-vegetarian critical theory.* New York, NY: Continuum.

Berlatsky, N. (2015). *Wonder Woman: Bondage and feminism in the Marston/Peter Comics,* 1941–1948. New Brunswick, NJ: Rutgers University Press.

Collings, S., Parson, S., & Arrigoni, A. (2014). Until all are free: Total liberation through revolutionary decolonization, groundless solidarity, and a relationship framework. In A. J. Nocella II, J. Sorenson, K. Socha, & A. Matsuoka (Eds.), *Defining critical animal studies: An intersectional social justice approach for liberation.* New York, NY: Peter Lang.

Costello, M. J. (2009). *Secret identity crisis: Comic books and the unmasking of Cold War America.* New York, NY: Continuum.

Dittmer, J. (2013). *Captain America and the nationalist superhero: Metaphors, narratives, and geopolitics.* Philadelphia, PA: Temple University Press.

Eco, U., & Chilton, N. (1972). The myth of Superman. *Diacritics, 2*(1), 14.

Fawaz, R. (2016). *The new mutants: Superheroes and the radical imagination of American comics.* New York, NY: New York University Press.

Frymer, B., Van Heertum, R., Nocella, A., II, & Kashani, T. (Eds.). (2010). *Hollywood's exploited: Public pedagogy, corporate movies, and cultural crisis.* New York, NY: Palgrave Macmillan.

Giroux, H. A. (2015). *Zombie politics and culture in the age of casino capitalism.* New York, NY: Peter Lang.

hooks, b. (2004). *We real cool: Black men and masculinity.* New York, NY: Routledge.

Jones, P. (2005, October). What's wrong with rights? Retrieved from http://www.satyamag.com/oct05/jones.html

Kahn, R. (2010). *Critical pedagogy, ecoliteracy, and planetary crisis: The ecopedagogy movement.* New York, NY: Peter Lang, Inc.

Nocella, A., II. (2007). Unmasking the Animal Liberation Front using critical pedagogy: Seeing the ALF for who they really are. *Journal for Critical Animal Studies, 5*(1), 1–10.

Reclus, E. (1901). "On vegetarianism." Retrieved from http://www.ivu.org/history/europe19b/reclus.html

Rousseau, J., & Cress, D. A. (1992). *Discourse on the origin of inequality.* Indianapolis, IN: Hackett Publishing Company.

Saunders, B. (2011). *Do the gods wear capes? Spirituality, fantasy, and superheroes.* London: Continuum.

Singer, P. (1975). *Animal liberation: A new ethics for our treatment of animals.* New York, NY: New York Review.

Socha, K. (2012). *Women, destruction, and the avant-garde: A paradigm for animal liberation.* Amsterdam: Rodopi.

Wertham, F. (1954). *Seduction of the innocent.* New York, NY: Rinehart.

Wright, B. W. (2001). *Comic book nation: The transformation of youth culture in America.* Baltimore, MD: Johns Hopkins University Press.

# *Part III: Educational Industrial Complex*

# 5. *Teaching to End Human Supremacy↔Learning to Recognize Equity in All Species*

JOHN LUPINACCI

The stakes are high, and the capacity of the planet for sustaining life depends upon future generations learning to live in harmony and at peace with the diverse ecosystems within which they reside. In this chapter, I set out dialogically to invite you—the reader—into a conversation of sorts that seeks to mimic the pedagogical process I enact each time I teach. An activist-teacher from Detroit—now rooted with my family in Eastern Washington, I am an equity scholar-activist-educator. I work in teacher education where I teach folks who work, or will work, in schools or their communities to critically and ethically consider the importance of a culturally relevant and empowering education for every child. Whether I'm teaching a course on elementary social studies methods or courses on sociology of education and cultural studies in education, I do so with explicit objectives that students learn to rethink—and in many cases (un)learn—the role anthropocentrism (and the closely associated human-male-white-ableist supremacy) plays in how they understand and value relationships—especially teaching and learning. With strong influences from anarchism, critical pedagogy, and an ecojustice framework, I work to value the voices and lived-experiences of each student in all my teaching; but it's far from "anything goes." I invite students to think differently and to consider, and value, the different thinking of their peers. Furthermore, in all my teaching I invite students and colleagues to consider voice, agency, and value in the more-the-human members of the communities to which we all belong. I'll share insight into how I do that in this chapter and do so in a similar way that I teach—through dialogue. A dialogue through which I welcome future educators and educational researchers to consider a life as an activist-scholar committed to addressing domination in all its diverse forms. In the spirit of such a pedagogical dialogue, this chapter is not solely an academic critique of anthropocentrism, rather I invite us (you the reader and I) together to work to challenge dominant worldviews perpetuating unjust suffering and seek ways of engaging educators and educational researchers in doing the same. In this chapter, I am drawing from ecocritical projects in education—including critical animal studies (Best, 2009; Nocella II, Sorensen, Socha, & Matsuoka, 2014; Twine, 2010); anarchism (Amster, DeLeon, Fernandez, Nocella, & Shannon, 2009; Nocella, White, & Cudworth, 2015; Haworth, 2013; Suissa, 2010) and ecofeminism (Code, 2007; Plumwood, 1993, 2002; Warren, 1990, 2000)—all the while recognizing that there are centuries of wisdom in Indigenous epistemologies not explicitly included in this chapter. As a teacher, steeping in privileges associated with Western industrial culture and often teaching students who also come from Western culture, I focus on what Foucault (2010) referred to as an ontology of ourselves. I also aim to turn the critical and ethical lens inward while recognizing that the position I take as a teacher educator and educational researcher is privileged because I am a member of Western industrial culture interested in engaging in how "we" can together cease to do such harm and learn to do better for one another and the more-than-human world. The work in this chapter is not aimed at those whose narratives and knowledges embody an ecological worldview but rather focuses on the majority of educators and educational researchers with whom I encounter regularly as we—and they—go about business-as-usual without critically examining their—and our—implicit and explicit roles in reproducing injustice in the world. In sum, I will share a bit about a pedagogical process aimed at helping educators to recognize an anthropocentric worldview, to examine how this worldview is implicated in maintaining human (and male, white, able-bodied) supremacy, and to rethink—or (un)learn anthropocentrism in favor of ecological alternatives that are socially just and encompass all living systems.

As a teacher educator and an activist-scholar influenced and inspired by critical animal studies, I am deeply committed to a critical animal pedagogy's powerful assertion for educators to call into question ourselves as human beings in relationship to social suffering and environmental degradation. I am not alone in this line of inquiry as evidenced by the growing body of critical animal studies scholarship and the emergence of radical animal studies. Furthermore, the work from many of those sharing their work in this book represents the diverse bodies of scholarship and stories that a multitude of critical scholars are drawing from to rethink dominant discourses and discursive practices influencing power and current regimes of oppressive ideologies.

All of the influences contributing to this work recognize that there are powerful phenomena occurring within schools and in a society that are profoundly abusive and are violently reproducing relationships that make racism, sexism, classism, ableism, and speciesism, to name a few, seem inevitable and inescapable. For many of us teaching through a critical animal pedagogical framework we perceive that, to be in school, by situation of its location in society, means learning to function with/in, accept, and submit to the authority of a tremendously exploitive culture. This so-called *reality* is preparing young people for their fate in a very unhealthy and, in many ways, broken society. Despite enduring such an assimilating and horrific training for so many of us and our children, there are existing relationships that offer not only an alternative but a plethora of alternatives. In life, or in the living systems within which humans exist, there are alternative examples of the all too often rationalized abuse. There is story after story about how living together ought to be, about what it could be, and about who and what belongs to and revels together in what we might call our communities but more accurately refer to as our living-local ecologies. These are the stories, smells, tastes, joys, and pains that explore the potential of community and the power of a decentralized locally sovereign existence. These are the stories of activism whispered, sung, barked, danced, chirped, carved, painted, howled, and in all cases offer alternatives to what is and unapologetically welcomes the possibility of a very different society. For many students and educators committed to an interspecies equity, this is a call to action to identify the role education both plays and ought to play in transitioning toward socially just and sustainable communities. The call asks us to directly confront and overcome human supremacy—to recognize how, in Western industrial culture, we make meaning through a human-centered hierarchical worldview and to recognize how such a worldview manifests in our daily relationships.

It is rare that we share enough stories in the classroom about how people take action in defense of animals but it would be totally human-centric of us to not share a few stories about how the more-than-human world resists exploitation and captivity, and elicits survival. One such story, that I typically introduce in my courses to teach how human supremacy is played out, and how one might learn from acts of animal resistance, is that of Ken Allen. In zoos all over the world, primates—more specifically chimpanzees and orangutans, have been observed and recognized for their elaborate escape plans and protests. Ken Allen, nicknamed "the hairy Houdini" over his long stay in the San Diego Zoo where he was born and died of lymphoma in 2000, found countless ways to escape his confinement (Silver, 2013). Ken would repeatedly be found wandering the zoo peacefully strolling the premises. While zookeepers and trainers knew Ken took to unscrewing any screw he could find and removing nuts from bolts, they couldn't figure out how he was getting loose. It was such a mystery to them that they installed video surveillance. However, Ken knew about the surveillance and avoided escaping on camera. Trainers dressing up as visitors to try and monitor him also never fooled him. Ken repeatedly escaped and found ways to find cover and get out of confinement. Gaining such fame brought an intrusive local news crew to come and do a feature on Ken to which he greeted them by pelting them with his feces. Over and over, the zoo spent excessive money and called upon long lists of experts and couldn't contain Ken. Even the punishment didn't stop Ken from escaping. Ken Allen would mislead trainers, find and hide tools, and defy all odds to stroll the zoo freely (Hribal, 2008). While Ken Allen may be one of the most famous escape artists, he is not alone, nor is he the first or last.

The pedagogical moral of this story shouldn't be about the intelligence of primates but rather that of general animal resistance. Jason Hribal (2010), a historian that researches and documents animal resistance, in *Fear of the Animal Planet: The Hidden History of Animal Resistance* shares a wide array of defiant acts that have been observed ranging from tigers, previously injured by poachers only to turn around and hunt down their assailants to kill them, and only them, leaving other humans in the villages unscathed. Then there are elephants who after repeatedly being stabbed and injured by handlers using a bullhook finally decide to stomp out their abuser (Hribal, 2010). While I am not condoning or celebrating the killing, I do want to highlight that these animals are aware of the situation they are in and in most cases do not resort to violence. However, it is clear that all animals have power and when pushed will exercise that power intentionally. Why, though, share these examples in classrooms? I share the stories of animal resistance and intelligence because I want students to consider that diverse species have agency, thinking skills, and experience emotions—especially, to debunk the myth that human beings are the only species capable of planning ahead. Furthermore, while students expect a course committed to social justice and sustainability to include activism, but rarely do the students come to class considering the ways animals and other species like insects, plants, fungi, and bacteria (to name a few) engage in resistance. It may seem commonsensical to many reading this chapter, but for many of my students—the future teachers and school leaders of North American K–12 schools—they often express that they have never considered animals and other species for that matter, as members of the community with equivocal value and capacity for resistance. Teaching these stories early on in the semester sets up a foundational base upon which I introduce my political views in favor of total liberation. It also becomes an entry point for my sharing the principles of critical animal studies and my anarchist (anti-authoritarian and anti-oppression) orientation toward teaching and learning. Later in this chapter, I'll share how I build on this foundation not only to consider agency, emotion, resistance, and intelligence among more-than-humans, but also to consider their worth as equivocal to human beings—in other words, how in my teaching I work to identify and confront human supremacy in how we understand ourselves in relationship to other beings.

When thinking about the diverse work among scholars and activists challenging human supremacy, it is through recognizing, respecting, and representing our solidarity in a growing movement that I consider how human supremacy works. Whether it be through factory farmed food, clear cutting forests, fracking the land, or dumping plastic into the oceans, one needn't venture far in modern industrial society to witness any myriad of unjust suffering endured by both humans and more-than-human beings. Given the depth at which human supremacy operates, it is paramount that teachers learn to recognize how such assumptions of superiority not only operate in our day-to-day lives, but also how it simultaneously works to dominate our perceptions of self, one another, and who is constituted as a member of an often romanticized, but important concept—community. In the following section, I will share stories, goals, admissions, confessions, and dreams with students and colleagues in ways that I hope complement, inspire, and challenge current perceptions of the roles teachers play in Western industrial schools—especially in universities and within the realm of teacher preparation. With the shared goals of social justice and sustainability in the form of truly decentralized communities, when we listen to the voices of the members who are silenced in our day-to-day lives we grow to learn things. These are things that we might wonder how we could have ignored and often continue to struggle with ignoring—things which we grow to know too much about and to which in life we simply owe too much to let continue. In listening to the voices of each other, our animal kin (like Ken Allen) and the life systems to which we all belong we can begin to learn together with a commitment to the well-being of all species to consider what it might entail putting into practice interspecies equity.

After setting up the foundational basis for students to consider interspecies equity, I then commit three (sometimes four or five) class meetings to examining how social justice and environmental sustainability are deeply entangled within Western industrial cultural assumptions that everything is to be commodified, colonized, and dominated. As a human being in Western industrial culture, this sort of day-to-day systemic violence is predicated on the dominant assumption that such an understanding of the world is culturally constructed and dependent upon the notion of a human being as an individual, separate and superior to all other beings. In education courses, it's not uncommon for students to come to class with a strong interest in the potential role schools might play in alleviating social suffering. However, what many college students don't often know coming into the field of education is how much of a role schools have played, and continue to play, in creating and maintaining inequality. This is why I teach the relationship between authoritarianism, patriarchy, and capitalism and the North American prison industrial complex (Alexander, 2012; Davis & Barsamian, 1999), neoliberalism (Harvey, 2007; Lupinacci & Happel, 2015), and the animal industrial complex (Adams, 1997; Noske, 1989; Twine, 2012) together with environmental degradation and the school-to-prison pipeline (Nocella, Ducre, & Lupinacci, 2017) in my education courses. Given that students may or may not be aware of the interconnectedness of multiple systems of oppression, I introduce the "10 Principles of Critical Animal Studies" developed by Steven Best, Anthony J. Nocella II, Richard Kahn, Carol Gigliotti, and Lisa Kremmerer in 2007 (Nocella, Sorenson, Socha, & Matsuoka, 2014, pp. xxvii–xxviii) and emphasize how critical animal studies activist-scholar educators argue for a critical praxis that:

> advances a holistic understanding of the commonality of oppressions, such that speciesism, sexism, racism, ableism, statism, classism, militarism and other hierarchical ideologies and institutions are viewed as parts of a larger, interlocking, global system of domination. (p. xxvii)

In a course I recently taught titled "Globalization, Identity, and Education," I structured an opening for a dialogue about commonalities among forms of oppression. I shared with students that we are living/teaching at a time in which:

- 700 million people are living in extreme poverty. (United Nations, 2014)
- The United States has only 5% of the world's population but 25% of the world's prisoners (NAACP, 2015). Furthermore, in US prisons, African-Americans and Latinos/Latinas accounted for 58% of all prisoners in 2008 (NAACP, 2015). And, as many as 50% of all prisoners in the United States are diagnosed with having a mental illness or another type of disability. (National Disability Rights Network, 2012)

- Since 1970, there has been a 52% species loss and 76% loss of fresh-water species. (WWF, 2014)
- Global emissions of $CO_2$ have increased by 50% since 1990. (United Nations, 2014)

The common thread woven through these staggering statistics is a political and economic system of exploitation in global systems of capitalism made possible by assumptions of human superiority that expand to the far reaches of the planet. How could such exploitation be rationalized, justified, or ignored? How can we learn from the imprisonment and resistance of our more-than-human kin like Ken Allen? Following the tenth principal of CAS, we as critical educators can, and should, seek "to create openings for constructive critical dialogue on issue relevant to Critical Animal Studies" (Best et al. in Nocella et al., 2014, p. xxvii). In other words, we as critical educators can teach, how is it that we challenge dominant assumptions about who/what benefits and who/what suffers? Pursuing these questions, I am working with other scholar-activists engaged in critiquing such assumptions of Western industrial culture with the following primary assertions we often refer to as an ecocritical framework.

In many of the courses I teach, we read very closely ecofeminist philosopher Val Plumwood because of how her work connects the unjust suffering inflicted upon women with the subjugation and destruction of nature in patriarchal cultures. Ecofeminists claim that there is an overlying logic that informs all domination. Warren (2000) calls this logic the "logic of domination," and she and other ecofeminists understand this as informing all relationships that are based on hierarchies. Within the logic of domination, hierarchical binaries inform thinking and ways of being in the world. In conversation with and similar to CAS activist-scholarship, ecofeminists insist that binary/dualistic thinking about both women and the environment inevitably lead to the devaluation of women, nature, and nonhuman beings. Ecofeminists, therefore like CAS activist-scholars advocating for work that "deconstructs and reconstructs the socially constructed binary oppositions between human and nonhuman animals" (Best et al. in Nocella et al., 2014, p. xxviii), are concerned with critiquing and deconstructing binary thinking and the logic of domination that accompanies it.

Since those oppressed in modern society are in almost all cases feminized and naturalized, ecofeminist philosopher Val Plumwood (1993) suggests that through ecological feminism, we can perceive how value-hierarchized dualisms—superior/inferior dualisms like culture/nature, reason/emotion, mind/body, and man/woman—work discursively to marginalize women, other human groups, nonhuman animals, and nature. Plumwood maintains that these dualisms are inseparable from each other and work to form the root discourses—such as anthropocentrism (human-centrism) and androcentrism (male-centrism)—that create and recreate oppressive and exploitative relationships. She examines how forms of centric thinking work to exclude, homogenize, background, incorporate, and instrumentalize life to create what Karen Warren (1990) calls "a logic of domination" (p. 139). I work from this position in class with students to consider the ways diverse species of trees become lumber, animal flesh becomes meat, and children become prisoners. As students are reading and discussing Plumwood's (2002) logic structure for centric thinking, we focus in on the analytical tools she provides for deconstructing domination. Centric thinking sets up one term, group, or individual as "the One" primary center and defines anything outside of the center as marginalized "Others" (p. 101). Plumwood's logic structure for centric thinking includes: a) Radical Exclusion: The Other is excluded and deemed both inferior and separate from the central position/one; b) Homogenization/Stereotyping: The Otherized group is seen as an undifferentiated whole; c) Denial/Backgrounding: The Other, already hyperseparated and grouped as inferior, becomes the target of the center "to represent as inessential" (p. 104). Further, since the center is dependent on the Other, in order to uphold the logic of centrism, such dependencies are denied; d) Incorporation: The assimilation of the Other into the center; e) Instrumentalism: The assimilation of the Other as a colonized resource.

An ecocritical framework draws on and builds from key ecofeminist claims. Sparing the details of the development of an ecopedagogy (Kahn, 2010) and an EcoJustice educational framework (Martusewicz, Edmundson, & Lupinacci, 2015), I summarize it on three interrelated fronts. In an ecocritical framework, activist-scholar educators strive to identify the role that education both plays and ought to play in transitioning toward creating diverse, socially just, and sustainable communities. Drawing from an EcoJustice educational framework (Martusewicz et al., 2015) and contributing to the growing field of ecocritical work in social and cultural foundations of education, I have together framed this with feminist qualitative methodologist Dr. Alison Happel-Parkins as an approach to teaching and learning as critically and ethically

1. examining Western industrial culture and its impacts on social and environmental systems;
2. examining the links and interrelations between prevalent value-hierarchized dualisms and how superior/inferior dualisms intertwined with

either/or thinking prevalent in Western industrial culture contributes to inequities such as racism, classism, sexism, ableism, anthropocentrism, and so on; and

3. examining and identifying how to teach or share skills and habits of mind that support socially just and environmentally sustainable communities. (Lupinacci & Happel-Parkins 2016, p. 41)

So, as such scholar-activist educators, we turn our attention toward the discourses, and associated discursive practices, that manifest as the logic structures of centric thinking that constitute Western industrial culture. Specifically, we focus our attention on what CAS scholars refer to as a "holistic understanding of the commonalty oppressions" (Nocella et al., 2014, p. xxvii). Referring to the entanglement of speciesism, sexism, racism, ableism, classism, and other value-hierarchized ways of thinking as I teach, I acknowledge all of these and their associated practices, emphasizing how they are learned, and potentially confronted and unlearned in schools if teachers commit to combating the complexly interwoven role that undergird all forms of oppression. While not prioritizing any form of oppression over another, I focus on going deeply into how anthropocentrism works in Western industrial culture to undergird the rationalizing of subjugating and dominating people, animals, the land, oceans, and skies.

## *Anthropocentrism*

While several dominant discourses shape and get shaped by mainstream—or man-stream, white-stream, and human-stream—culture in my teaching, I focus heavily on one of these modernist discourses—anthropocentrism, or human-centered thinking as it contributes to the gangland of human supremacy. I choose anthropocentrism with the conviction that in most cases when examining social justice and sustainability it exists unchecked. It provides a foundation for injustices directly experienced by other modernist discourses. However, I am not trying to argue that it supersedes other dominant, modernist discourses, as they all exist in a complex relationship to constitute modernity, or what might be considered by sociologist Zygmunt Bauman as a late modernity culture (Bauman, 2000; Martusewicz et al., 2015). While teaching toward learning to end human supremacy, it is important to facilitate dialogues that openly discuss our general observations of anthropocentric influences on daily life—especially in teacher education. First, before returning to opening up in a dialogue about human supremacy—similar to the story about Ken Allen at the start of the chapter—to explain what I teach in my education courses, consider the story of the elder elephant crisis.

## *The Elder Elephant Crisis!*

In Africa poaching and massive habitat loss devastate elephant communities; as a result, they additionally suffer from a loss of elder elephant knowledge. Over the past 100 years, the elephant population on the continent of Africa has dropped dramatically. The World Wildlife Fund (WWF) reports that 3–5 million elephants were roaming Africa in the 1930s and 1940s but that due to hunting for trophies and tusks in the 1950s elephant populations plummeted. The WWF reports: "In the 1980s…an estimated 100,000 elephants were being killed per year and up to 80% of herds were lost in some regions. In Kenya, the population plummeted by 85% between 1973 and 1989" (http://wwf.panda.org/what_we_do/endangered_species/elephants/african_elephants/).

While recent efforts to protect the population have seen some success, what has been observed is an increase in hyper-aggression, depression, and unpredictable behaviors. What is sad, but fascinating is that scientists are reporting changes in communication among this nonhuman species due to a loss of elder elephants in the elephant communities. In an essay titled Elephant Breakdown (2005), Bradshaw, Schore, Brown, Poole, and Moss explain that elephants are renowned for their close social relationships. Further, Bradshaw et al. (2005) report that "male socialization begins during infancy with the mother and a tight constellation of all mothers. But in adolescence, males leave the natal family to participate in older all-male groups" (p. 807). They explain that the elder males in elephant societies play a critical role in the socialization of adolescent male elephants to quell aggression and learn to be adult elephants. Similarly, female elephants separated from elder female elephants aren't socialized to how to raise young elephants. Bradshaw et al. (2005) explain: "Young elephants are reared in a matriarchal society, embedded in complex layers of extended family. Culls and illegal poaching have fragmented these patterns of social attachment by eliminating the supportive stratum of the matriarch and older female caretakers" (p. 807).

What this highlights is that without elders to teach young elephants how to live and be contributing members of strong, healthy elephant communities, future generations of elephants are susceptible to dangerous behaviors. This situation perpetuates a cycle in which the elephants are further separated from one another and from the habitat in which they would mostly likely thrive together as social beings. This example teaches us that humans are not the only species teaching our young through a complex communicative system of language that passes on important knowledge on how to survive together. Further, we are reminded of the importance of elder knowledge. We are not the only species with teachers. Following a CAS pedagogy, this example shares with students in my courses—who are themselves very closely linked to identities of being both a teacher and a student—an important link that reminds us that we humans can in fact learn from the diverse species among us. We read theorists and philosophers too often in education courses without ever recognizing the knowledge systems that reside among more-than-human relationships. In other words, through a CAS pedagogy in my teaching I focus dialogues on what we might learn about teaching from reading Val Plumwood, Richard Kahn, and Helena Pederson and considering what we are also learning from elephants and other more-than-humans all around us.

In the spirit of continuing this chapter through dialogue, I thought it appropriate to share a version of a letter I often use as pedagogical practice in classrooms and research. The following is an open letter I share with my friends, colleagues, students, future friends, and of course to you the reader. I also emphasize that while often dominant schooling imposes that the desired result is that we agree, the intent here is to engage in a dialectic in which often differing ideas are brought together. In the spirit of diversity—a condition of difference upon which all things exist, I think it's important to emphasize we don't all agree. In fact, it is in recognizing those differences that we can talk authentically about solidarity and how we may together work toward alleviating the injustices we find ourselves drawn toward dismantling. My intention is to share a pedagogical tool, which some might consider taboo, to vocalize how I teach through a critical animal pedagogy. There is value in committing through our teaching to not hide such projects. We should muster the will to say it out loud more often, recognizing that anthropocentrism is often so unchecked in our institutions and has material implications in the current moment that it has to be dealt with sensitively. In addition to addressing anthropocentrism, this particular pedagogical strategy of letter writing is intended to foster the fire of an antihuman supremacist charge in solidarity with the fire of social justice and sustainability. In other words, this work explores the ways in which we value difference to teach in solidarity of total liberation. The next section is a letter I share with students and colleagues, and then I encourage students to write letters to their students, future students, and to their teachers—both the human and more-than-human. It is my hope that before this chapter ends you find yourselves (you the reader) in a dialogue with the text and inclined to write letters of your own that explore and examine your relationships with diverse species, equity, equality, and education.

### *A Letter to Spark an Anti-Human Supremacist Charge in Education*

Dear Teachers, Scholars, and Activists in Western Industrial Culture:

I see anthropocentrism enacted every day, all of the time, and I need not venture far to be reminded of its broad acceptance—even among some of the most socially conscious and environmentally concerned advocates of change. Anthropocentrism is an unspoken taboo among those who fancy themselves as advocates of justice. To name the ways in which this deeply embedded discourse of modernity plays out requires that I speak outside of the frameworks of what is conventionally recognized among my peers as reasonable. When I bring specific attention to the injustices done to our more-than-human kin, I often hear that I cannot be serious. I am.

I have come to terms with the fact that for some of my human pals, this discourse will generate heated contention and that I am willing to sacrifice my social standing in the name of the billions murdered represents a measure of my rationality. This letter pays homage to the graves of so many nameless kin, some of whom are humans, reduced to objects, as their torture and deaths are rationalized simply because they are likened to animals—and therefore expendable and exploitable as resources of human supremacy. We are so quick to dismiss the value of another species as inferior to our existence because we are told they do not feel pain or they are without a soul. What I hope to teach through my letter is the understanding that when we disregard the value of a nonhuman life, we are vulnerable to extending that condition of value onto one another.

I am not ashamed, nor shy, to say that this so-called *logic* or *reasoning* is often "a conscience eased by lies" (Hannah, 1996). Anthropocentrism justifies the unjust in the name of holding up the status of human as a supreme being. It cannot be ethically denied that the preservation of humans as superior beings is a justification for suffering in the form of domination, exploitation, and murder. Yet, this is often the case. And any resistance to this position is all too often dismissed as unrealistic or irrational.

I am not ashamed to say that I too slip into, and live, daily in hypocrisy as I find myself deeply entrenched in anthropocentric discursive practices and living life in accordance to an illusion of disembeddedness—as though I exist as an individual. Going about my day fancying uttered meanings, forgetting that beneath all those metaphors language reminds us of our dependencies. Gregory Bateson in *From Versailles to Cybernetics* (1972) describes how our language, at its roots, reveals our dependency on one another and the ecosystems to which we belong. He writes, "Mammals in general, and we among them, care extremely, not about episodes, but about the patterns of their relationships" (1972, p. 478). As humans, we are all animals and akin to other species. We all communicate through that kinship—our shared dependency—that we interpret experiences by recognizing or being attuned to patterns of our ecological existence. Bateson (1972) goes on to explain:

> When you open the refrigerator door and the cat comes up and makes certain sounds, she is not talking about liver or milk, though you may know very well that that is what she wants. You may be able to guess correctly and give her that—if there is any in the refrigerator. What she actually says is something about the relationship between herself and you. If you translated her message into words, it would be something like, "dependency, dependency, dependency." She is talking, in fact, about a rather abstract pattern within a relationship. From that assertion of a pattern, you are expected to go from the general to the specific—to deduce "milk" or "liver". This is crucial. This is what mammals, animals, are all about. They are concerned with patterns of relationship, with where they stand in love, hate, respect, dependency, trust, and similar abstractions, vis-à-vis somebody else. (p. 478)

Bateson directs our attention to language to emphasize the historical importance of events and experiences that shape how we interpret patterns. Simply put, language shapes how we understand and then act in the world. We, of course, use signs and symbols to communicate these patterns of our relationships, but beneath all of the layers of signs and symbols is dependency, dependency, dependency.

So, any conversation or communication reminds us that beneath the cultural context is the ecological reality that none of us exist independent from anything. We are all interconnected. Following a core tenant of CAS—interdependency, our interconnectedness reminds us of our ecological existence. Together, by recognizing and valuing interconnectedness we are reminded of our dependency. With every breath we take and utterance made, we are communicating dependency, dependency, dependency.

This means that we are members of a shared place—an ecological system. In this system, we make sense of patterns based on past experienced patterns passed onto us from our parents, teachers, peers, and anyone with whom we experience socialization or any relationship for that matter. So, we end up with metaphors that we use to shape how we move from the abstract to the specific. How funny is it to think that we are all running, or walking through life, yapping away with all sorts of deep meaning in what we say and what is consistently hidden is the ecological translation—dependency, dependency, dependency.

It is troubling how in our culture we have come to devalue dependency as a weakness or as something undesirable. We have rejected this matter of fact and replaced it with an illusion of autonomy. This is what Bateson is referring to when he talks about the illusion of humans being separate from their dependency on everything else, and Plumwood when she calls this the "Illusion of Disembeddedness."

So as a committed ecocritical educator—and more importantly as a member of an ecological system to which we are akin to and dependent upon as living members—I am forced to confront the vast sea of indifference to the suffering and slaughtering of our kin with all my heart, mind, and body. As a person drawn toward confronting social injustices, I came into the field of education with a burning desire to work toward confronting and challenging the unjust suffering perpetuated by schools and society. I worked, and continue to work, ardently to recognize how the privileges of race, class, gender, sexual orientation, and ability contribute to oppression. I recognize how these forms of social injustice exploit and marginalize, but when I fail to consider equity among so-called *other species*, I neglect to recognize a key assumption in all the aforementioned injustices. I miss the root of these inequalities because all around me, including within my own frame of reference, we are under the influence of anthropocentric discourse. So to take this discourse into account, simply put, life is too short to unjustly cause another's to be shorter. If I believe that humans are superior beings, then in order to cast others as inferior, I need only to liken them to nonhuman species. This is the position from which many of my friends and adversaries argue and sharply informs the following question:

To my very *brilliant* and *productive* friends, why do you ignore the issue and deny the clear relationship between anthropocentrism, consumerism, and brutality? When I speak of our responsibility to all members of our community, you roll your eyes and, intentionally or unintentionally, your dismissal marginalizes, socially isolates, and preys on the shame of those who would otherwise be willing to respond to the suffering of those in pain. While in the grasps of anthropocentrism, you cringe at the idea of being equal in worth to another species, yet you speak that you live toward equality for all humans. This is an issue of acting out of selfishness and superiority, and before you even listen you find yourself disciplined by your impulse to reject the worth of any other species as equivocal to that of your own.

I thought that I shared common ground with advocates of social justice, and I do as long as together we challenge and confront the conventions fed to us by a past and present language that interprets living, breathing, feeling beings as mechanized objects. While among company who don't challenge anthropocentrism, it's only a matter of time until any one of us is served as your commodity. The institutionalized violence and oppression of racism, slavery, the subjugation and objectification of women, the genocide on Native Americans, and the mass incarceration of Black and Brown bodies, and so on…all hinge on reasoning that the other is less in value and is not fully human. Do you still feign indignant to injustice? Because without addressing anthropocentrism—or conversely, your willingness to uphold your human superiority—you are, in fact, making way for the rape, slavery, and murder of the human and the more-than-human family: the animals, the land, each other, and the oceans to which we all belong.

## *Conclusion: Learning from More-Than-Human Teachers*

The more I address anthropocentrism among social justice advocates, the more I recognize the tendency of the privileged to dismiss what many would rather not confront. On the one hand, while I admire, value, and am a firm supporter of a shared commitment to respond to the undeniable atrocities that we—as Western industrial human beings—enact on one another, none of these atrocities occur in isolation. It is paramount that we work as allies to those suffering while challenging and confronting the systemic roots of oppression on our respective fronts. Through critical animal pedagogy we have a responsibility, as privileged members of society, to support the oppressed in whatever capacity we can. Together our goal is to break the will of the oppressor.

In my courses, I invite students to identify or seek out a more-than-human teacher for the semester and to write this teacher a letter expressing their interest in learning from what this teacher is willing to offer. After writing this letter I ask the students to journal weekly what they learn from this *different* sort of teacher–student relationship. Often students write about how different and challenging the assignment may be; however, after a few weeks the discussions deepen. A few weeks into the assignment while we are addressing how we envision and are planning to teach for social justice and sustainability, students share about what they are learning from their more-than-human relationships. The stories shared are rich and provide an excellent context for discussing how it is we are learning to teach from participating in a different sort of teaching/learning relationship. Simultaneously, we are practicing listening to the more-than-human world and learning together how to respect diverse species and their intrinsic worth and value in our living and local ecologies—our community. We write about sitting in silence and what they hear, smell, and feel. Further, I ask students to consider what they are learning and how that learning process is going specifically in relationship to social justice and sustainability. In efforts to build their abilities to share such often shamed relationships, students provide feedback to each other and report out to each other on what it is they are learning. The results are exciting conversations and sharing of stories as well as the students teaching each other as they share their stories. As I share stories about crows that teach each other to fabricate and use tools (Kenward, Rutz, Weir, & Kacelnik, 2006), tales from the adventures of Ken Allen, fungal wonders of mycelial networks, and connect these stories to our everyday lives, I ask how is it we continue to treat one another and the more-than-human world with such disregard, disrespect, and cruelty. We often say, let us learn from our more-than-human teachers but how often do we allow ourselves to confront our human-supremacist illusions of superiority and autonomy. In the end, the point is that we learn together to listen to the many wise voices we often don't hear and certainly don't consider as our teachers. It's always amazing to share stories and then engage in critically and ethically examining our assumptions and the lessons we are learning. It comes down to the simple project that a critical animal pedagogy is relational learning often between and among diverse beings and about what we learn when we humble ourselves and learn to listen to the river, the wind, the maple tree, the osprey, the wolves, the worms…and the list goes on.

## *References*

Adams, C. J. (1997). "Mad cow" disease and the animal industrial complex: An ecofeminist analysis. *Organization & Environment, 10*(1), 26–51.

Alexander, M. (2012). *The new Jim Crow: Mass incarceration in the age of colorblindness.* New York, NY: The New Press.

Amster, R., DeLeon, A. P., Fernandez, L. A., Nocella, A. J., & Shannon, D. (2009). *Contemporary anarchist studies: An introductory anthology of anarchy in the academy.* New York, NY: Routledge.

Bauman, Z. (2000). *Liquid modernity.* Malden, MA: Polity Press.

Best, S. (2009). The rise of critical animal studies: Putting theory into action and animal liberation into higher education. *Journal for Critical Animal Studies, 7*(1), 9–52.

Bradshaw, G. A., Schore, A. N., Brown, J. L., Poole, J. H., & Moss, C. J. (2005). Elephant breakdown: Social trauma: early disruption of attachment can affect the physiology, behaviour and culture of animals and humans over generations. *Nature, 433*, 807.

Code, L. (2007). *Ecological thinking: The politics of epistemic location. Studies in feminist philosophy.* New York, NY: Oxford University Press.

Davis, A. Y., & Barsamian, D. (1999). *The prison industrial complex.* Chicago, CA: AK Press.

Foucault, M. (2010). What is enlightenment? In P. Rabinow (Ed.), *The Foucault reader* (pp. 32–50). New York, NY: Vintage Books.

Hannah, C. (1996). Nailing Descartes to the wall/(liquid) meat is still murder [Recorded by Propagandhi]. On *Less Talk, More Rock* [Audio Cassette]. San Francisco, CA: Fat Wreck Chords.

Harvey, D. (2007). *A brief history of neoliberalism.* New York, NY: Oxford University Press.

Haworth, R. H. (2012). *Anarchist pedagogies: Collective actions, theories, and critical reflections on education.* Oakland, CA: PM Press.

Hribal, J. (2008). Orangutans, resistance and the zoo. Counterpunch. Retrieved from http://www.counterpunch.org/2008/12/16/orangutans-resistance-and-the-zoo/

Hribal, J. (2010). *Fear of the animal planet: The hidden history of animal resistance.* Petrolia, CA: CounterPunch.

Kahn, R. (2010). *Critical pedagogy, ecoliteracy and planetary crisis: The ecopedagogy movement.* New York, NY: Peter Lang.

Kenward, B., Rutz, C., Weir, A., & Kacelnik, A. (2006). Development of tool use in New Caledonian crows: inherited action patterns and social influence. *Animal Behaviour, 72*, 1329–1343.

Lupinacci, J., & Happel, A. (2015). Recognize, resist, and reconstitute: An ecojustice response to neoliberalism. In M. Abendroth & B. Porfilio (Eds.), *Schools against the neoliberal rule* (Vol. 1, pp. 269–287). Charlotte, NC: Information Age Press.

Lupinacci, J., & Happel-Parkins, A. (2016). Recognize, resist, & reconstitute: An eco-critical conceptual framework. In J. Diem (Ed.), *The social & cultural foundation of education: A reader* (pp. 34–56). San Diego, CA: Cognella.

Martusewicz, R., Edmundson, J., & Lupinacci, J. (2015). *EcoJustice education: Toward diverse, democratic, and sustainable communities* (2nd ed.). New York, NY: Routledge.

National Association for the Advancement of Colored People (NAACP). (2015). *Criminal justice fact sheet.* Retrieved from http://www.naacp.org/pages/criminal-justice-fact-sheet

National Disability Rights Network. (2012). *Criminal justice.* Retrieved from http://www.ndrn.org/issues/criminal-justice.html

Nocella, A. J., II, Ducre, K. A., & Lupinacci, J. (2017). *Addressing environmental justice and dismantling the school-to-prison pipeline: Poisoning and incarcerating youth.* New York, NY: Palgrave.

Nocella, A. J., II, Sorensen, J., Socha, K., & Matsuoka, A. (2014). *Defining critical animal studies: An intersectional social justice approach for liberation.* New York, NY: Peter Lang Publishing.

Nocella, A. J., II, White, R. J., & Cudworth, E. (2015). *Anarchism and animal liberation: Essays on complementary elements of total liberation.* Jefferson, NC: McFarland & Company, Publishers.

Noske, B. (1989). *Human and other animals.* London, UK: Pluto Press.

Plumwood, V. (1993). *Feminism and the mastery of nature*. London, UK: Routledge.

Plumwood, V. (2002). *Environmental culture: The ecological crisis of reason*. New York, NY: Routledge.

Silver, A. (2013). Run for it: Ken the orangutan, San Diego. In TIME's *Top 11 Zoo Escapes*. Retrieved from http://content.time.com/time/specials/packages/article/0,28804,2041628_2041646_2041609,00.html

Suissa, J. (2010). *Anarchism and education: A philosophical perspective*. Oakland, CA: PM Press.

Twine, R. (2010). *Animals as biotechnology: Ethics, sustainability and critical animal studies*. New York, NY: Routledge.

Twine, R. (2012). Revealing the 'animal-industrial complex'—A concept & method for critical animal studies? *Journal for Critical Animal Studies*, *10*(1), 12–39.

United Nations. (2014). *The millennium development goals report 2014*. Retrieved from http://www.un.org/millenniumgoals/2014%20MDG%20report/MDG%202014%20English%20web.pdf

Warren, K. (1990). The power and the promise of ecological feminism. *Environmental Ethics*, *12*(2), 125–146.

Warren, K. (2000). *Ecofeminist philosophy: A Western perspective on what it is and why it matters*. Lanham, MD: Rowman & Littlefield Publishers.

WWF. (2014). *Living planet report 2014: Species and spaces, people and places*. In R. McLellan, L. Iyengar, B. Jeffries, & N. Oerlemans (Eds.), *WWF*. Gland, Switzerland. Retrieved from http://wwf.panda.org/about_our_earth/all_publications/living_planet_report/

# 6. *Intersecting Oppressions: The Animal Industrial Complex and the Educational Industrial Complex*

MENEKA REPKA

## *Introduction*

This chapter was inspired by a past issue of the *Journal for Critical Animal Studies* (Thomas & Shields, 2012), which troubled the links between prisons and animals. I am interested in further expanding this parallel to also include the examination of institutions of education as places where nonhumans and humans intersect. Though many of these ideas could easily extend to post-secondary institutions, I focus here on K-12 (Kindergarten to 12th grade) education in Alberta (Canada), as this is where the bulk of my experiences are drawn from. In this chapter, I understand the animal industrial complex (AIC) as triple helix of influential, powerful systems that control knowledge systems about meat production: government, the corporate sphere, and the academy (Twine, 2012, p. 17). The concept of an AIC was originally put forth by Noske (1989) who proposed that the commodification of nonhuman animals in food systems is directly linked to capitalist systems that prioritize "monopolistically inclined financial interests" over the well-being of humans, nonhumans, and the environment (p. 22). Twine (2012) expands on Noske's work, explaining that "corporate influences have had a direct interest through marketing, advertising, and pecia manipulation in constructing the consumption of animal products as a sensual material pleasure" (p. 15). Similarly, the educational industrial complex revolves around ideology, profit, and technology as the three central components that organizations and agencies use to generate revenue while undermining real scholarship and learning (Picciano & Spring, 2013). As a teacher and activist, I have long been concerned with the assumption that schools are neutral places. As well, I want to open a dialogue about how both human and nonhuman animals collide with forces of authority and dominance in the meat industry and the industry of public schooling. As institutions that are products of particular cultures, schools tend to mirror the dominant discourses of the societies in which they are situated. In this chapter, I am particularly interested in the dominance of a carnistic ideology. Joy (2010) defines carnism as a sub-ideology of speciesism, where the classification of nonhuman animals into "edible" and "nonedible" categories reflects a deeper belief system (pp. 28–29). Joy (2010) explains that the terms "meat eater" (a dietary orientation) or even "omnivore" (a biological orientation) are inaccurate because they assume that "the practice of consuming meat [is] divorced from a person's beliefs and values. [These terms imply] that the person who eats meat is acting outside of a belief system" (p. 29). Although vegans and vegetarians are typically thought to be the only people who bring their beliefs to the table, all people who participate in an industrial society are behaving according to beliefs. Importantly, carnism is a violent ideology because "it is literally organized around physical violence...if we were to remove the violence from the system—to stop killing animals—the system would cease to exist" (Joy, 2010, p. 33). Entrenched, violent systems, such as carnism, avoid scrutiny through invisibility: they remain unnamed and their violent practices are sequestered and hidden from the public (pp. 30–35). I begin my examination of carnistic discourses in schools by observing the explicit ways in which the slaughterhouse is eerily similar to the school; these disturbing resemblances reflect a larger social dependence upon the commodification and suppression of that which threatens to disrupt the interests of dominant individuals or groups. I then turn to the school cafeteria and science lab to examine the more complex and nuanced ways that schools are intertwined and involved with the factory farm. I engage with the messy layers of codependence that have worked alongside capitalism and permitted both systems to thrive. Finally, I theorize that the prevalence of speciesist narratives in schools facilitates the conditions for discrimination and leads to the possibility of the school to be considered a concentrated site of multiple, interrelated oppressions. Schools are incredibly diverse places, but the industrial obsession with homogeneity provides greater opportunities for discrimination rather than collaboration. Though a blueprint for flawless schooling is likely unattainable, I propose that a radical reconsideration and transformation of both the systems of industrial food production and education must occur. Neither system can exist without the other to reinforce it, so the same radical change, an elimination of hegemonic (Gramsci, 1971) capitalist ideologies, is the first step toward a more just and peaceful society.

## *Systems for Children and Systems for Nonhumans*

There are some expected resemblances between the educational industrial complex (Picciano & Spring, 2013) and the AIC (Noske, 1989; Shukin, 2009; Vialles, 1994). Structurally, both types of complex rely on surveillance and disciplinary power to maintain conformity, productivity, and efficiency. Schools are informed by an assembly-line approach to learning, which also mirrors systems of food production. In both cases, large populations of beings spend a great deal of time concentrated within concrete buildings, and move through a series of stages in production before they are considered ready for society to consume; both children and nonhuman "food" animals are treated as goods rather than as individuals, and the ultimate aim of both systems is conformity. Both students and nonhuman farm animals are segregated into compartments or rooms and are further mechanized by lining up for a variety of procedures (such as eating or going outside) throughout the day. As well, both students and nonhuman animals are organized quantifiably by being assigned a number (i.e., an ear tag or a student identification number or card). If either an animal or a student slows down the efficiency of the system, they are either medicated or removed. For example, in 2011, 11% of school-aged children were diagnosed with ADHD and 6% of those students were receiving medication (Visser et al., 2014); it is also common for schools to segregate students with behavioral or learning differences. Both public education systems and factory farming systems operate with the interests of corporate investors in mind. One example is the frequency with which schools are allied with specific companies that provide discounted or subsidized textbooks, computers, computer programs, and cafeteria food (Molnar, 2002). This corporate funding can exasperate disparity between schools, especially when funds are unevenly distributed in favor of particular subjects or programs, such as school sports teams that receive sponsorships. Industrial farms also have stakeholders. Livestock genetics companies, animal consultants, and agricultural or animal science programs at universities all benefit from the commodification of nonhuman animals (Scully, 2002). The structural similarities between the AIC and educational industrial complex signify an institutional reliance upon routinization, depersonalization, and commodification in order to prevent these powerful systems from collapsing and losing their positions of dominance.

## *From the Slaughterhouse to the Schoolhouse*

In addition to the mirroring of schools and slaughterhouses through their literal architecture and institutional practices, contemporary institutions of public education have also developed processes that reinforce and perpetuate the meat and related industries. Simultaneously, industries that benefit from the oppression of nonhumans prioritize education as a means to further entrench and solidify their rhetoric in young generations of students. This process of reciprocal value sharing between institutions is enacted through the various forms of curriculum to covertly normalize nonhuman exploitation. The perpetuation of an anthropocentric discourse occurs through specific teaching resources that are made abundantly convenient to teachers, the literal consumption of nonhumans through school cafeterias, and the objectification of nonhuman bodies through the practice of dissection. These three interrelated domains represent the predominant ways in which factory farms are overlapped and entangled with public schooling.

## *Understanding Curriculum*

In this section, I feel it would be prudent to contextualize the types of curriculum that are constantly in tension within formal learning environments. While a multitude of curricular facets have been established in curriculum theory (Eisner, 1985; Ornstein & Hunkins, 2009), I draw up on Eisner (1985) to broadly classify curricular discourse within either the explicit, implicit, or null curriculum (p. 103). The explicit curriculum refers to what teachers, administers, policy makers, and governments intend for students to learn through classroom activities and materials. The prescribed curriculum is an example of the explicit curriculum; it refers to the written, formal, planned aspects of learning that typically originate from a professional organization or government. The implicit or hidden curriculum describes the unintended values that are transmitted to students through the structure and policies of the school. For instance, students may implicitly learn that remaining passive and compliant will yield praise from authority figures, even though such values are not necessarily stated overtly. The null curriculum refers to the intentional omission of either entire disciplines or specific pieces of information and prevents or restricts teaching opportunities for these topics; by avoiding particular subjects, schools send the message that these areas are unimportant. Eisner (1984) proposes that the null curriculum can prevent critical thinking skills in students by reinforcing a parochial perspective (p. 97) rather than "a wide range of modes of thought" (p. 98). In most cases, all three types of curriculum operate on some level.

## *Teaching Resources*

The explicit curriculum is directly targeted by the meat, dairy, and egg industry as a method of ensuring that messages are delivered through the perceived authority of curricular documents. For notoriously overworked public school teachers, lesson plans that are advertised as correlating with provincial or state mandates for learning can be a welcome relief from the time-consuming and often tedious work of developing daily lesson plans. Though the meat, dairy, and egg industries design curricula and lesson plans that are meant to teach students about nonhuman animal agriculture, these resources are at best biased, and at worst factually incorrect. In a document designed for teachers by the Alberta Farm Animal Care (AFAC, 2009), "beak trimming" is described as a necessity that makes it easier for chickens to eat (p. 11). Aside from reinforcing a paternalistic rhetoric of "care" and a speciesist positioning of the human as superior to the bird, the lesson plan does not allude to the primary reason that debeaking occurs: when chickens are kept in tight spaces, they begin to peck one another to death and ultimately decrease profits. The stress of beak amputation on the chicken, though documented in literature (Gentle, 1990; Hofstad, 1984), is also not discussed in the teacher resource. In Alberta, "Aggie Days" (Agriculture Days) is another venue for propagating a single dominant narrative about nonhuman animals. According to the Aggie Days website, teachers can bring their elementary students to a full-day expo, where children "learn more about Alberta's agricultural and historical legacy" (Calgary Stampede Agriculture, 2015, para. 3). In an anecdotal account of one Aggie Days exhibit, a visitor recalls: "I remember from the exhibit at Stampede that there is a life size cow with a life like vagina…they get kids to line up, stand on a stool, and put their arms in, violating the cow to 'feel her fetus!'" (Telena, personal interview, 2015). In other words, students actually learn that the forcible impregnating of a female cow is not only acceptable, but a learning opportunity as well. Implicitly, such activities also reinforce a social reluctance to universally accept consent culture (across genders, sexualities, and species), as well as deny the necessity for bodily autonomy to be applicable to all bodies. Therefore, by infiltrating schools through curricular texts, nonhuman animal agriculture industries work alongside schools to promote a violent ideology under the guise of learning.

### *The Cafeteria*

Like other hegemonic systems, the AIC and the educational industrial complex become "mutually reinforcing systems of domination" (Nibert, 2014, p. x). Even if a teacher chooses not to utilize the resources provided by the meat, dairy, and egg industries, the mythologies of carnistic ideology surreptitiously permeate the daily interactions between students and the content of their learning. This is evident in the school cafeteria, where the consumption of animal flesh and secretions is normalized and promoted. For instance, children are encouraged to drink cow's milk largely through school lunch programs and advertising campaigns. A great deal of money goes into marketing research that determines how details such as packaging, flavoring, and temperature can increase the consumption of cow's milk in school-age children and youths (Campbell & Campbell, 2006, p. 293). The advertising of cow's milk occurs through a self-referential process, where the dairy industry is constantly portrayed as important and necessary. School cafeterias and breakfast programs operate in tandem not only with lesson plans but also facets of public pedagogy such as the Disney Channel, Treehouse, Nickelodeon, and the Cartoon Network (Campbell & Campbell, 2006, p. 293), which all feed the same messages about the benefits of milk into one another. The authoritative circulation of dominant narratives about food is further carried out through the distribution of food under the assumption that it is universally consumable. There are strong colonial implications of circulating dairy products in schools that are ethnically diverse. The majority of people of color and Indigenous populations are lactose intolerant (Ornelas, 2014) and do not have a tradition of cow's milk consumption. Cow's milk, and its associations with health, embodies a Eurocentric conception of nutrition that is not compatible with the global norm. School milk programs have historically been attached to the notion of perfection, connecting the development of children with the development of entire nations. The inherent "perfection" of cow's milk was emblematic not only of the supposed nutritional qualities of the drink, but also of the perfectability of bodies and the superiority of specific bodies (DuPuis, 2002). This positions bodies that cannot produce lactose as "deficient" and implicitly implies that children who do not adopt Western dietary norms are weaker (DuPuis, Harrison, & Goodman, 2011).

Because Albertan school curricula beyond elementary school are generally anthropocentric, meat in the cafeteria is one of the only places where junior high or high school students encounter nonhuman animals in schools. Providing sustenance for one's body warrants deep reflection about how the privileging of the human species is essentially what drives the process of nonhuman animals showing up on the lunch tray as food. The consumption of nonhuman animal flesh presents a valuable opportunity for "understanding and relating oneself and all other beings" (Weitzenfeld & Joy, 2014, p. 25). Yet the counternarrative that these conversations may provide, like well-balanced vegan cafeteria options, are marginalized and suppressed by a dominant carnistic discourse (Weitzenfeld & Joy, 2014, p. 25). By isolating and segregating students from an engaged and reflexive understanding of their own food, the null curriculum (content that formal curricula avoid) is censored, surveilled, and monitored by the hegemonic interests of particular groups and institutions. The school cafeteria also functions as a disciplinary space, where specific mechanisms function to order and discipline specific bodies (DeLeon, 2011; Foucault, 1975). The cafeteria space, like spaces of industrial meat production, forces individuals into processing lines, where unique personal identities are quashed in favor of the same Western and capitalist values that permeate classrooms, such as efficiency and conformity. As DeLeon (2011) observes, the positioning of students as units who must militaristically grab a lunch tray while waiting in line demonstrates how the gaze of the state uses particular spaces to dominate and oppress (pp. 185–188). Feeding nonhuman animals to students also reproduces human supremacist understandings of how various species should be ordered and categorized. In the cafeteria, the bodies of "farm" nonhuman animals are of import only insofar as their capacity to provide food for human children. The literal "meat" that students consume is only an abstraction of its animal origin, and this "invisibility renders the [animal] subject meaningless" (DeLeon, 2011, p. 188). In contrast, nonhuman animal bodies employ an elevated status in science classrooms, as they perpetuate the myth of human progress as a result of nonhuman oppression.

## *The Science Lab*

In addition to the use of nonhumans as food in school cafeterias, the human/nonhuman animal binary is reinforced in science classrooms. The positioning of nonhuman animals as research tools in the process of dissection operates within the larger structural and systemic framework of socially accepted violence to segregate the nonhuman as "other." For example, dissection was originally introduced to public school curricula in the 1920s (Orlans, 1993) to mirror the field of science as a whole, and 75%–80% of students will dissect at least one nonhuman animal during their K–12 school tenure (Balcombe, 2003). Aside from the obvious environmental complications and potential physical harm to the animals themselves (Oakley, 2009), the social and pedagogical implications of dissection call for critical reflection and evaluation. As Oakley (2009) points out, the "attitudes, values, and beliefs" (p. 61) that are generated by the null curriculum can function to orient nonhuman animals as tools of research rather than as beings that should be included in human circles of moral and ethical consideration. The practice of dissection elevates the status of humans by feeding the speciesist perspective that the most salient value of a nonhuman animal lies in their capabilities to satisfy human desires or curiosities. Like nonhuman animals used in vivisection labs at a postsecondary level, the covert message of dissection in junior high and high school is that nonhuman animals are objects, commodities, and curios to be disembodied and reduced to their anatomical components rather than valued inherently as beings that exist for their own purposes. In the quest to transform schools into spaces that not only promote peace and justice, but actually demonstrate and fully engage in peaceful and just practices, it is clear that serious contemplation about the ways in which nonhuman animals are entwined with public education is needed. What kind of world can our youth create if curriculum shifts away from anthropocentric and speciesist narratives and instead compassionately embraces the nonhuman world as valuable regardless of its capacity to serve humans?

## *The School as Inherently but Unintentionally Intersectional*

As an institution of cultural reproduction (Bourdieu, 1973, 1974, 1979, 1991; Dewey, 1902; Pedersen, 2010), public schooling absorbs and disseminates socially dominant ideologies that have gradually become entrenched as widely accepted and generally unexamined truths. The persistence of both implicit and explicit speciesist discourses in K–12 schools results in perhaps the most violent and sinister lesson of all: that it is desirable for nonhumans, because they are less socially powerful, to be exploited for human interests. This is a dangerously precarious message that highlights speciesism as an early and socially accepted form of discrimination. A salient implication of speciesist narratives being circulated and perpetuated through educational discourse is the repositioning of the school as a site that is fruitful for interrelated oppressions to culminate. The capitalist and industrial model of schooling nourishes not only the values of speciesism, but also sexism, racism, ableism, classism, and discrimination toward those in the LGBTQIA+ community (as well as other oppressions). This inequitable climate is cultivated through conformity, competition, and single-issue activism rather than total liberation efforts in schools.

## *Conformity in Schools*

North American public schooling is rooted in the Prussian-industrial model of education, where the primary goal of learning was "to enslave, and not to enfranchise, the human mind" (Chubberley, 1920, p. 488). Of industrial schooling, Meschaninov (2012) writes:

> Children were depersonalized and isolated from each other at an early age. Seated in rows, they were easily silenced, controlled, and forced to engage in rote tasks whose sole purpose was to inculcate obedience. Taught fragmented subjects that deprived them of context and perspective, their thinking was intentionally and systematically stunted. (p. 4)

The persistence of this historically inherited school culture is evidenced through the continued fragmentation of curriculum into subject disciplines, rigorous standardized testing, and individualism. As an elaboration of the factory assembly line, the privileging of conformity in today's schools undermines the unique knowledge, experience, and perspectives of students who do not fit the dominant Western viewpoint of what success or education must mean. For instance, test scores do not measure kindness, empathy, imagination, critical thinking, and collaborative skills. The industrial emphasis on social homogeneity also means that their peers further marginalize students who do not fit into the preestablished criteria of "in-groups." As Kosciw, Greytak, Palmer, and Boesen (2014) note, 74% of students who identify as gender or sexual minorities are called names or threatened because of their sexual orientation. It is to be expected that a hierarchical system that implicitly values conformity would lead to the marginalization of not only nonhuman animals, but also humans who do not fit the criteria of dominant groups.

## *Competition in Schools*

The emphasis on individual test scores or grades is also problematic. The spirit of competition that was initially designed to differentiate future leaders from obedient workers (Meschaninov, 2012, p. 8) further exasperates the magnifying oppressions faced by many students and quashes opportunities for collaboration. Because the value of schools is largely influenced by test scores, individual subjects are ordered hierarchically based on how they relate to standardized tests: math and hard sciences are near the top, and the arts at the bottom. This hierarchy is reflected both economically and socially, as arts programs are notoriously underfunded, and are usually the first to be eliminated or scaled back. Leonhard (1991) for instance found that 60% of schools surveyed budgeted $500.00 or less for music programing and 67% of schools budgeted $1000.00 or less for visual art. This prioritization of subjects also functions to solidify future social roles, as students who perform highly on tests are assumed to be better suited to specialized careers with greater monetary gains. The competitive atmosphere in schools also places value on the notion of hierarchy and creates interrelated binary systems: academically strong/weak, physically strong/weak, racial majority/racial minority, human/nonhuman. If it is acceptable for humans to dominate nonhumans, and it is acceptable for the academically strong to be praised while others are not acknowledged, the conditions are dangerously suitable for many further dualisms to follow.

### *Where Are Total Liberation Efforts in Schools?*

In addition to fragmenting and decontextualizing subject areas, social justice groups in schools also seem to exist separately and in isolation, rather than collaboratively toward unified liberation. Schools operate as social microcosms to manage and redirect various types of oppressions, pacifying students rather than promoting meaningful social change. In an attempt to curb the massively discriminatory obstacles that many students of nondominant populations face, actors (including peers, teachers, principals, and other administrators) within the educational-industrial complex are encouraged to form extracurricular modes of cultivating social justice for each of these groups. This results in a smorgasbord of clubs or groups, such as the Black and Hispanic Student Alliance, the Gay-Straight Alliance, and Students for Environmental Awareness (to name just a few). Without minimizing the benefits of school support in giving voice and agency to previously powerless groups (Marshall & Scribner, 1991, p. 3), the fragmentation of these groups can also generate tensions that should be examined. First, students who identify as part of a minority group that is not highly represented at their school can be further marginalized or isolated. Rarely, for instance, are such clubs present for students who identify as disabled. More discouragingly, schools actually "face a disincentive to provide well for [special needs] students," as they fear grades will be pulled down and greater financial resources will be required (Brighouse, 2002, p. 7). As well, the presence of some social justice groups, such as a nonhuman animal or environmental rights group, can place the school in a contradictory and hypocritical position. This is evident in cases where schools promote nonhuman animal and environmental ethics through student groups, but continue to serve meat or engage in the dissection of nonhuman animals. Finally, the divisions between activist groups in schools can operate to reinforce solidarity only within individual groups, rather than collectively as well. This results in a further distancing of connections and entanglements between oppressions. By promoting single-issue activism, schools avoid the intersectional complexities of students who experience multiple discriminatory obstacles and prevent the possibility of total liberation. As Firzgerald and Pellow note, "we must enlist the support of a diverse cross section of humanity to embrace social justice in all forms" (2014, p. 36). The existence of social justice groups in schools does not address the fundamental problems of capitalist systems of education. Though educational policy and practice may change on a superficial level, the capitalist industrial system of schooling, along with historically rooted dominant ideologies, remains intact. For deep social transformation through the lens of total liberation, the intertwined links between oppressed groups needs to be fully understood by both students and society to more effectively challenge institutional hierarchical systems.

## *A Word on Public Pedagogy*

I must acknowledge that not all learning occurs within the constraints of a traditional classroom. As Drew and Socha (2015) note, the privileging of institutional education systems only reinforces political and economic power dynamics (p. 164). However, the values represented by the cannon of institutional learning seem to spill into other public institutions where human and nonhuman relationships are further complicated. In detailing the parallel between the AIC and the educational industrial complex, it is evident that instances of public pedagogy, such as zoos, aquariums, and circuses, are not exempt from capitalist and industrial models of learning. These institutions occupy spaces of both education and entertainment, and again emphasize human/nonhuman binaries. Building on the aforementioned examples in more formal conceptualizations of education in K–12 school, I focus my discussion in this section of the chapter on zoos, primarily because they are regarded as important educational and conservation facilities. Furthermore, such public pedagogies often play an intimate and underexamined role in formal curriculum. Trips to the zoo, for example, are widely heralded by teachers, administrators, parents, and students as extensions of classroom learning that serve to further enhance student understandings about nonhuman animals. Though usually geographically distanced from both schools and slaughterhouses, zoos (excluding those that primarily exist as sanctuaries) remain entangled in the same capitalist ideologues of efficiency, hierarchy, and commodification. Like slaughterhouses and traditional classrooms, zoos are designed to maximize their own interests in profit and growth. Patrons of the zoo move through a series of exhibits that has been specifically designed to generate traffic and structure the viewing experience of humans (Fogelberg, 2014). Zoos clearly denote hierarchy by mirroring the authoritative structure of a traditional classroom and by positioning humans as observers of nonhuman otherness. The assumption of both traditional schooling and zoos as instances of public pedagogy is that learning must be guided by an external expert. In zoos, employees or information signs take on pedagogical roles by providing information to visitors about the various nonhuman animals. As Fogelberg (2014) observes, [zoo signs] aptly represent institutional power at play: [...] they are penned by an anonymous "author" in a public institution (zoo) which is perceived as an "authority" figure by those that visit (whether implicitly or explicitly). This "authority" is furthered by the lack of references, resulting in an asymmetry of knowledge, information, and access... (p. 793).

Additionally, zoos create clear boundaries between the human and nonhuman world. Physically, nonhumans and their "environments" are segregated by the walls of their enclosures, highlighting their status as objects to serve human desires. This imagery closely resembles the packaging of nonhuman animal flesh behind a glass meat counter and the removal of nonhumans from their environmental contexts in classroom dissections. The commodification of nonhumans in zoos is further communicated by a zoological gaze, where the "status of the animal body [is] accessible for manipulation and control" (Pedersen, 2010, p. 59). As Pedersen (2010) notes during her field studies, "the desire for interaction, to get a *response* from the animal observed, was a driving force in many students' visual involvement with animals" (p. 59). Additionally, zoos mimic the capitalist model of traditional schooling by inviting corporate sponsorship opportunities. In the same way that schools supplement funding for programs through the promotion of corporations, zoos seek to mesh education with entertainment through trained nonhuman animal performances, "zoovenir" shops, and popular culture themes such as cartoons or movie characters (Pedersen, 2010, pp. 86–87). In addition to exasperating the objectification of nonhumans, these gimmicks remain problematic from a pedagogical standpoint as well because they assume that students will only be engaged and interested in subject matter that is directly tied to consumerist imagery. Finally, zoos as sites of public pedagogy mirror the sycophantic relationship between traditional public schools and the nonhuman agribusiness industry. Restaurants such as McDonalds make their way into zoos for the purpose of maximizing profits for both businesses, but are also situated in the bizarre paradoxical tension of nonhuman animal conservation and nonhuman animal exploitation. As Pedersen (2010) points out, McDonalds' entire business concept is dependent upon exploiting nonhumans and the industries of food, schooling, and public pedagogy (i.e., zoos) that operate in tandem to reinforce habits of food socialization (p. 88). Similarly, corporate partnerships between public schools and such fast-food giants as McDonalds perpetuate the commodification of education through public schooling. As schools become increasingly desperate for funding, corporate sponsorship programs capitalize on opportunities to circulate branding through not only cafeteria and vending machine food, but also school supplies, sports teams, promotional material tied into curricula, and academic incentives such as prizes for reading (Molnar, 2002). Again, rewarding students with gift cards toward fast food companies is pedagogically irresponsible, as the positive behavior will likely cease when the reward system ends. Furthermore, the notion that learning must be followed by an extrinsic reward implies that learning alone cannot be a positive experience. Likewise, positioning food as a reward that must be earned obscures the purpose of food as nourishment and discourages mindfulness about one's own hunger and fullness cues. In some cities, fast food restaurants are also purposely clustered within walking distance of public schools (Austin, Melly, Sanchez, Aarti, & Buka, 2005), visually reminding students that associations of fun and happiness are located in relation to products rather than learning. Therefore, I maintain that even sites of public pedagogy that seem removed from traditional institutions of education are caught up in anthropocentric, Eurocentric, and capitalist values of education, as well as corporate interests tied to profit accumulation and branding. These embedded values obfuscate authentic and holistic learning and reinforce oppressive systems. Critical Animal Pedagogy (CAP) seeks to provide alternative ways of knowing that can challenge and resist this discourse.

## *The Role of CAP*

The links between industrial schooling and industrial processing of nonhuman animals for food are of particular relevance in CAP, which seeks to dismantle both of these processes by eliminating capitalism and the interwoven deep cultural habits of patriarchy, ethnocentrism, and anthropocentrism. CAP calls for a radical shift in systems of education, where hierarchy and competition are stigmatized rather than accepted as the norm or peddled as necessary. Capitalist models of education thrive by powerfully linking institutional structures of schooling with informal spaces of learning and for-profit industries. This is evident through the physical structure of these establishments, as well as their values of efficiency, hierarchy, and concentrated power. Thus, systems of education become social scripts that "[legitimize] the human domination of nonhuman nature" (Bell & Russell, 2000, p. 190). Conversely, CAP is interested in challenging relationships of oppression, privilege, and power in education (Bell & Russell, 2000; Lupinacci, 2013, 2015; Lupinacci & Happel-Parkins, 2015). CAP seeks to decentralize positions of absolute authority by uniting educators, scholars, community members, and the more-than-human (Abram, 1996) world in a framework of solidarity and ethical responsibility. CAP draws upon an interdisciplinary range of fields to establish a pedagogy that dismantles the status quo (Kahn, 2010) and seeks peace and justice as necessary components of life and learning.

The field of CAP is uniquely positioned at the intersections of critical pedagogy, anarchist models of education, ecopedagogy, intersectional radical activism, and humane education. Through integrating ideas from these models, CAS offers an important counterdiscourse that questions the oppressive humanist assumptions and relationships that are developed through typical institutes of formal learning. According to Nocella II (2004), one of the most salient elements in critical pedagogy is the need to break down the subject–object barrier to disrupt the fallacy of objectivity (p. 197). Critical pedagogues remain intimately entangled with the subjects they seek to better understand, and acknowledge their own interpretive positions (Freire, 1997; Nocella II, 2004, p. 198). In classrooms, this means that students and teachers relinquish their formal titles and strive to come to understandings collaboratively. This disruption of hierarchy is also an application of anarchist pedagogy. Anarchist theory questions the assumption that "institutions remain central to the system" (Drew & Socha, 2015, p. 165), arguing in favor of radical and liberatory change rather than reformist methods that rely on government. An anarchist approach to schooling may mean unlearning patterns of social conditioning (desks, lines, raising hands) that are prevalent in traditional school structures. In a CAP approach to learning, these petty mechanisms (Foucault, 1975) of control would be replaced with communities of learning that center upon intergenerational sharing of knowledge and recognizing the relationships between humans, nonhumans, and the Earth (Lupinacci, 2015; Lupinacci & Happel-Parkins, 2015). Drawing upon Indigenous epistemologies and ecopedagogy, CAP rejects the anthropocentrism of Western industrial schooling by encouraging an ethic of care for all species and the planet. Kahn (2010) elaborates on Freire's (1997) conception of critical pedagogy as a movement to resist authority and confront injustice to include meaningful change against environmental wrongs committed by humans (pp. 4–7). These ideas are especially needed when some environmental education programs do not even mention nonhumans as worthy of consideration within a study of the environment (Thirukkumaran, 2016). CAP is also influenced by humane education, an approach to learning that asks students to critically examine their daily actions with the goal of rejecting systems that are unsustainable or unjust (Weil, 2004). In line with anarchist models of learning, humane methods of education mean that students are motivated by the desire to resolve problems rather than a grade or other extrinsic reward system. Humane education also encourages students to buy items that are produced ethically (i.e., without child labor or nonhuman suffering) in favor of conventional versions of those products. While responsible consumerism when possible is certainly a step in creating social change, those of us in the human community must maintain an awareness that buying even "ethical" and vegan products continues to uphold a larger capitalist and exploitative system. Notions of "voting with one's dollar," "consumer power," and other clichés are extensions of capitalistic strategies to promote unlimited growth and distract from the greater goal of dismantling larger power structures. In my view, CAP can play a significant role in inviting students into a dialogue about true collective liberation that is distanced from capitalism.

In addition to maintaining an interdisciplinary framework, CAP extends the principles of critical animal studies into pedagogical practice. This includes incorporating nonviolent resistance tactics developed by activist groups such as the Animal Liberation Front and Earth Liberation Front (Nocella II, Sorenson, Socha, & Matsuoka, 2014). In my experience as a classroom teacher, schools are notoriously authoritative and unwelcoming of counterdiscourses that challenge existing dominant perspectives (such as the human-centered attitudes adopted by curricular structures). I am proposing that engaging the alternative pedagogies encompassed by CAP should also include street activism as necessary for social change. Students who self-identify as being a part of (or in support of) a marginalized group should not limit their learning to only activities such as presentations or research reports to draw attention to important social issues. Rather, a CAP perspective would support students in organizing protests, demonstrations, rallies, marches, sit-ins, petitions, and other types of street activism both within their school and local communities. Similarly, CAP elaborates upon the Institute for Critical Animal Studies' position not to support the prison industrial complex, a system that works alongside other complexes to further marginalize specific groups and increase profits through incarceration. Schools are directly linked to the prison industrial complex because practices that typify public schooling in North America, such as high-stakes testing, police officers in schools, and punishment systems for youths and children, lead to higher dropout rates. When combined with other social conditions such as economic disadvantages and familial stressors, these practices contribute to higher rates of incarceration among young people through the school-to-prison pipeline (Heitzeg, 2009). Education built upon CAP principles would promote "avenues for transformative justice, a non-punitive justice process that addresses oppression, interpersonal injustices, and accountability" (Nocella II et al., 2014, p. xxx). Rather than focusing on consequences for student "misbehavior," CAP advocates working alongside students in a continuous dialogue about how concerns, frustrations, and injustices may evoke feelings of anger, and that students are a necessary voice in creating positive change.

## Conclusion

Though there are structural points of symmetry between the educational industrial complex and the AIC, the intimate links between the two complexes make them entwined in deeply entrenched ways. The persistence of a dominant speciesist discourse in curricular texts, school cafeterias, and science labs reinforces a hierarchy between humans and nonhumans. When combined with the climate of conformity and competition in schools, I theorize that this binary generalizes to make the hierarchies between human groups even more prevalent. Schools become junctures of concentrated intersecting oppressions because they are modeled to suppress and segregate nondominant perspectives. Though this results in the formation of various activist groups in schools, the collaboration and unification of these groups is not typically discussed. Therefore, most schools and spaces of public pedagogy that continue to follow an industrial model work against the possibility of total liberation. CAP provides a meaningful counterdiscourse to the dominant structures of education that continue to operate through a framework of capitalist ideology. As an interdisciplinary and holistic pedagogical model, CAP seeks to empower youths, educators, activists, and community members to resist injustice and work collaboratively toward positive social change.

## References

Abram, D. (1996). *The spell of the sensuous.* New York, NY: Random House.

Aggie Days. (2015). In Calgary Stampede Agriculture. Retrieved from http://ag.calgarystampede.com/events/aggie-days/

Alberta Farm Animal Care Association. (2009). Elementary. In *A teacher's resource on farm animal welfare in Alberta* (pp. 1–40). Calgary, Canada.

Austin, S. B., Melly, S. J., Sanchez, B. N., Aarti, P., & Buka, S. (2005). Clustering of fast-food restaurants in schools: A novel application of special statistics to the study of food environments. *American Journal of Public Health, 95*(9), 1575–1581.

Balcombe, J. (2003). Assessment of alternatives in education. In N. Jukes & M. Chiuia (Eds.), *From guinea pig to computer mouse: Alternative methods for a progressive, humane education* (2nd ed., pp. 40–53). Leicester, England: InterNiche.

Bell, A. C., & Russell, C. L. (2000). Beyond human, beyond words: Anthropocentrism, critical pedagogy, and the poststructuralist turn. *Canadian Journal of Education, 25*(3), 188–203.

Bourdieu, P. (1973). Cultural reproduction and social reproduction. In R. Brown (Ed.), *Knowledge, education and social change*. London: Tavistock.

Bourdieu, P. (1974). The school as a conservative force: scholastic and cultural inequalities. In J. Eggleston (Ed.), *Contemporary research in the sociology of education*. London: Methuen.

Bourdieu, P. (1979). Symbolic power. *Critique of Anthropology, 4*, 77–85.

Bourdieu, P. (1991). Epilogue: On the possibility of a field of world sociology. In P. Bourdieu & J. Coleman (Eds.), *Social theory for a changing society*. Boulder, CO: Westview Press.

Brighouse, H. (2002). *School choice and social justice* (pp. 6–7). Oxford Scholarship Online. Retrieved from ERIC (10.1093/0199257876.001.0001).

Campbell, T. C., & Campbell, T. M. (2006). *The China study: Startling implications for diet, weight loss and long-term health*. Dallas, TX: BenBella Books.

Chubberley, E. P. (1920). *Readings in the history of education: A collection of sources and readings to illustrate the development of educational practice, theory, and organization*. Boston, NY: Houghton Mifflin Company.

DeLeon, A. (2011). What's that nonhuman doing on your lunch tray? Disciplinary spaces, school cafeterias, and possibilities of resistance. In S. A. Robert & M. B. Weaver-Hightower (Eds.), *School food politics: The complex ecology of hunger and feeding in schools around the world* (pp. 183–200). New York, NY: Peter Lang.

Dewey, J. (1902). *The child and the curriculum: Including, the school and society*. New York, NY: Cosimo, Inc.

Drew, L., & Socha, K. (2015). Anarchy for educational praxis in the animal liberation movement in an era of capitalist triumphalism. In A. J. Nocella II, R. J. White, & E. Cudworth (Eds.), *Anarchism and animal liberation: Essays on complementary elements of total liberation* (pp. 163–177). Jefferson, NC: McFarland & Company, Publishers.

DuPuis, E. M. (2002). *Nature's Perfect Food: How milk became America's drink*. New York, NY: NYU Press.

DuPuis, E. M., Harrison, J. L., & Goodman, D. (2011). Just food? In A. H. Alkon & J. Agyeman (Eds.), *Cultivating food justice: Race, class, and sustainability* (pp. 284–307). Cambridge, MA: Massachusetts Institute of Technology.

Eisner, E. W. (1984). No easy answers: Joseph Schwab's contributions to curriculum. *Curriculum Inquiry, 14*(2), 201–210.

Eisner, E. W. (1985). Five basic orientations to the curriculum. In E. Eisner (Ed.), *The educational imagination: On the design and evaluation of school programs* (pp. 61–86). New York, NY: Macmillan Publishing.

Firzgerald, A., & Pellow, D. (2014). Ecological defense for animal liberation: A holistic understanding of the world. In A. Nocella II, J. Sorenson, K. Socha, & A. Matsuoka (Eds.), *Defining critical animal studies: An intersectional social justice approach for liberation* (pp. 28–48). New York, NY: Peter Lang Publishing.

Fogelberg, K. (2014, August 30). Unsilencing voices: A study of zoo signs and their language of authority. *Cultural Studies of Science Education*, *9*, 787–799. doi:10.1007/s11422-013-9566-8

Foucault, M. (1975). *Discipline and punish: The birth of the prison*. Paris: Editions Gallimard.

Freire, P. (1997). *Pedagogy of the oppressed*. New York, NY: Continuum.

Gentle, M. (1990). Behavioral evidence for persistent pain following partial beak amputation in chickens. *Applied Animal Behavioral Science*, *27*(1–2), 149–157.

Gramsci, A. (1971/2014). *Selections from the prison notebooks*. Q. Hoare & G. N. Smith (Eds. & Trans.). New York, NY: International Publishers.

Heitzeg, N. A. (2009). Heitzeg, N. A. (2009). Education or Incarceration: Zero Tolerance Policies and the School to Prison Pipeline. *In Forum on public policy online* (Vol. 2009, No. 2). Oxford Round Table. 406 West Florida Avenue, Urbana, IL 61801.

Hofstad, M. S. (Ed.). (1984). *Diseases of poultry* (8th ed.). Ames, IA: Blackwell Publishing.

Joy, M. (2010). *Why we love dogs, eat pigs, and wear cows: An introduction to carnism*. San Francisco, CA: Conari Press.

Kahn, R. (2010). *Critical pedagogy, ecoliteracy, and planetary crisis: The ecopedagogy movement*. New York, NY: Peter Lang Publishing.

Kosciw, J. G., Greytak, E. A., Palmer, N. A., & Boesen, M. J. (2014). *The 2013 National School Climate Survey: The experiences of lesbian, gay, bisexual and transgender youth in our nation's schools*. New York, NY: GLSEN.

Leonhard, C. (1991). *The status of arts education in American public schools* (pp. 25–28). Washington, DC: Department of Education. Retrieved from http://files.eric.ed.gov/fulltext/ED342708.pdf

Lupinacci, J. (2013). Eco-ethical environmental education: Critically and ethically examining our perceptions of being human. In A. Kulnieks, D. R. Longboat, & K. Young (Eds.), *Contemporary studies in environmental and indigenous pedagogies* (pp. 185–200). Rotterdam, The Netherlands: Sense Publishers.

Lupinacci, J. (2015). Recognizing human-supremacy: Interrupt, inspire and expose. In A. J. Nocella II, R. J. White, & E. Cudworth (Eds.), *Anarchism and animal liberation: Essays on complementary elements of total liberation* (pp. 179–193). Jefferson, NC: McFarland & Company, Publishers.

Lupinacci, J. & Happel-Parkins, A. (2015). (Un) Learning anthropocentrism: An eco-justice framework for teaching to resist human-supremacy in schools. In S. Rice &

A. G. Rud (Eds.), *The educational significance of human and nonhuman animal interactions: Blurring the species line* (pp. 13–30). New York, NY: Palgrave.
Marshall, C., & Scribner, J. (1991). *It's all political: Inquiry into the micropolitics of education. Education and Urban Society, 23*, 347–355.
Meschaninov, Y. (2012). *The Prussian-industrial history of public schooling* (pp. 1–10). The New American Academy.
Molnar, A. (2002). *Fifth annual report on commercialism in schools/Corporate branding of our schools* (pp. 1–10). Tempe, AZ: Commercialism in Education Research Unit, Education Policy Studies Laboratory, Arizona State University.
Nibert, D. (2014). Foreword, In A. Nocella II, J. Sorenson, K. Socha, & A. Matsuoka (Eds.), *Defining critical animal studies: An intersectional social justice approach for liberation* (pp. iv–xi). New York: Peter Lang Publishing.
Nocella, A. J., II. (2004). Understanding the ALF: From critical analysis to critical pedagogy. In S. Best & A. J. Nocella II (Eds.), *Terrorists or freedom fighters? Reflections on the liberation of animals* (pp. 195–201). New York, NY: Lantern books.
Nocella, A., II, Sorenson, J., Socha, K., & Matsuoka, A. (2014). Introduction: The emergence of critical animal studies. In A. J. Nocella II, J. Sorenson, K. Socha, & A. Matsuoka (Eds.), *Defining critical animal studies: An intersectional social justice approach for liberation* (pp. 110–134). New York, NY: Peter Lang Publishing.
Noske, B. (1989). *Beyond boundaries: Human and animals*. London: Pluto Press.
Oakley, J. (2009). Under the knife: Animal dissection as a contested school science activity. *Journal for Activist Science and Technology Education, 1*(2), 59–67.
Ornelas, L. (2014, December 20). Beyond veganism: Diet and consumption in a global world. In Earthling Liberation Kollective. Retrieved from https://humanrightsareanimalrights.com/2014/12/20/lauren-ornelas-beyond-veganism-diet-consumption-in-a-global-world/
Ornstein, A., & Hunkins, F. (2009). *Curriculum: Foundations, principles, and issues*. Boston, MA: Allyn & Bacon.
Pedersen, H. (2010). *Processes and strategies in human-animal education*. West Lafayette, IN: Purdue University Press.
Picciano, A., & Spring, J. (2013). *The great American educational-industrial complex: Ideology, technology, and profit*. New York, NY: Routledge.
Scully, M. (2002). *Dominion: The power of man, the suffering of animals, and the call to mercy*. New York, NY: St. Martin's Press.
Shukin, N. (2009). *Animal capital: Rendering life in biopolitical times*. Minneapolis, MN: University of Minnesota Press.
Thirukkumaran, M. (2016). *Where are the animals in environmental education curricula? A Discursive analysis. Green Theory and Praxis Journal, 9*(2), 73–86.
Thomas, S., & Shields, L. (Eds.). (2012). Prison and animals [special issue]. *Journal for Critical Animal Studies, 10*(2), 1–139.
Twine, R. (2012). Revealing the 'animal-industrial complex'—A concept & method for Critical Animal Studies? *Journal for Critical Animal Studies, 10*(1), 12–39.

Vialles, N. (1994). *Animal to edible.* Cambridge: Cambridge University Press.
Visser, S. N., Danielson, M. L., Bitkso, R. H., Holbrook, J. R., Kogan, M. D., Ghandour, R. M., & Perou, R. (2014, January). Trends in the parent-report of health care provider-diagnosed and medicated attention deficit/ hyperactivity disorder: United States, 2003–2011 [Electronic version]. *Journal of the American Academy of Child and Adolescent Psychiatry, 53*(1), 34–46. doi:10.1016/j.jaac.2013.09.001
Weil, Z. (2004). *The power and promise of humane education.* Gabriola, BC: New Society.
Weitzenfeld, A., & Joy, M. (2014). An overview of anthropocentrism, humanism, and speciesism in Critical Animal Theory. In A. Nocella II, J. Sorenson, K. Socha, & A. Matsuoka (Eds.), *Defining critical animal studies: An intersectional social justice approach for liberation* (pp. 3–27). New York, NY: Peter Lang Publishing.

# *Part IV: Lessons From the Classroom*

# 7. "We are one lesson": Some Reflections on Teaching Critical Animal Geographies Within the University

RICHARD J. WHITE

## *Introduction*

For activist-scholar educators aiming to raise consciousness around a radical intersectional politics, one that fully recognizes the interspecies struggle for social justice and total liberation, the university has long been an important site of struggle and contestation (Lupinacci, 2015; Nocella, Sorenson, Socha, & Matsuoka, 2014; Nocella, White, & Cudworth, 2015; Pellow, 2014). Drawing on some notable experiences rooted in my Human Geography module, *"Philosophy, Space and Place"* this chapter speaks of the possible transformative abilities that critical pedagogy/ies can bring within and beyond academia. Before doing so however, it is important that the significant pressures and vulnerabilities that activist-scholars experience within a higher education are not be overlooked. In this context, I wish to acknowledge the partiality of the situated sociospatial experiences that will be described in this chapter, filtered as they are through my own unearned privileges of being white, cis-male, heterosexual, and a senior academic at a UK university with a permanent contract. Despite this working within the university in ways that invites, for example, (1) critical reflection on everyday human–animal geographies rooted in speciesism, violence, and oppression; and (2) arguing for anarchist vegan geographies rooted in peace, social justice, and freedom has often proved an alienating and demoralizing experience. Indeed, agitating for social justice in an increasingly neoliberalized institution can also be tantamount to committing professional suicide (vis-à-vis tenure, academic progression, and so on). I'm sure for many critical animal scholars, while positive experiences in teaching critical animal studies in the university are a source of tremendous encouragement and satisfaction, these experiences are all too often set against a broader backdrop of negative conflicts where challenging speciesism, along with classism, racism, sexism, etc., engages a daily and demoralizing battlefield.

When reflecting critically on the last decade of lecturing in higher education, the majority of ugly skirmishes, and bruising exchanges, that I have personally encountered have almost exclusively involved my fellow professional colleagues. Indeed, adding insult to injury, some of the most disappointing examples of this have arisen when in dialogue with so-called "critical" or "radical" progressive scholars situated on the political Left, individuals who you may expect to find common expressions of support and solidarity (for commentary on the vilification of animals rights advocates across the political spectrum, both Left and Right, see Sorenson, 2011). To give one personal example: following a conference presentation on Critical Animal Geographies, an under-the-breath comment from the session chair was tossed in my direction: "You know, you *almost* made me feel guilty about that bacon I'd eaten this morning!" It certainly wasn't the most original or creative response that I had heard when talking about the rights of nonhuman animals to an intimate/hostile audience, which usually bounce between lazy accusations of "anthropomorphism" and occasionally toward the surreal: a senior colleague once asked, when we were exchanging views on the rights of nonhuman animals: "Why didn't you marry a dog?!" More commonly, hostile exchanges take the passive-aggressive nature of "only a joke", given, perhaps, as a form of defense by those who would rather not reflect on the immoral implications of their arguments, and face up to their culpability in supporting animal abuse: both human and nonhuman.

In-spite of these encounters, I believe that the university still offers critical animal scholars a tremendous site of hope, opportunity, and possibility for bringing the world we desire closer into view. I write this with some conviction, based on the critical pedagogies that I have sought to harness, with the aim of animating collaborative discussion with students (co-learners) whether that be in lectures, seminars, and/or informal conversations that take place elsewhere in/beyond the campus. Paying particularly close attention to my experiences of teaching critical animal geographies to undergraduate geography students, the aim of this chapter is to unpack *some* of the key pedagogical approaches used to promote a new radical imaginary around human–animal relations. To this end, as well as my personal reflections, the key themes and arguments draw on general conversations with students, formal module feedback, and more detailed one-to-one interviews. Through recognizing and valuing writing and teaching as meaningful examples of direct action, the modest hope is that—by reflecting on the experiences and approaches to teaching critical animal geographies—the chapter may well offer further ideas for others to constructively reflect on, and perhaps adapt and harness elsewhere.

In terms of structure, the focus of the chapter is principally an extended reflection on my experience of teaching—and the students' experiences of encountering—critical animal geographies within a module called *Philosophy, Space and Place (PSP)*. Broadly speaking there were three key components within these sessions that inform the structure here: (1) critical animal geography and moral thinking, (2) intersectionality, and (3) activist praxis. Having focused on these, the chapter then makes links with other important consequential developments, situated both within academia (focusing on final year Dissertation theses, for example) and beyond academic (recognizing the wider impact and pressures of students who have embraced a vegan praxis). Before continuing it is also worth highlighting that the *PSP* was replaced by a new module called Critical Geographies in 2017–18. This new module has taken "the best" of PSP, and continues to extend its commitment to embracing the ideas, vision, and praxis of total liberation, and push these within and beyond the university toward new, challenging, and radical directions.

## "We are one lesson": Promoting a Critical, Anarchist Imaginary around Human-Animal Relations

Between 2006 and 2016 *PSP* was consistently attended by around fifty predominantly white, British undergraduate human geography students enrolled in their second year at university. The students' classes were timetabled for an average of two hour sessions each week (up to twenty-four weeks in total per academic year). The module aimed to equip students with a detailed understanding of several theoretical perspectives that underpin approaches to geographical knowledge. A central emphasis of PSP focused on interrogating the radical and critical geographies that emerged within geography from the late 1960s and early 1970s, geographies which have influenced and continue to influence the discipline in the present day. While many critical geography syllabuses typically focus attention on Marxist and Feminist geographers, the module in question has always created meaningful space to explore anarchist praxis and critical animal studies. Anarchism has stressed the importance of critical pedagogy, and has been influential in informing much of the radical grounds in which critical animal studies finds itself (Drew & Socha, 2015; Kahn, 2003, Lupinacci, 2015, White, 2015a). This mutual dialogue between what is "critical" in animal geographies and the resurgence of anarchist praxis within dissident human geography is, arguably, the most vital, vigorous, exciting, and promising development within the dissident discipline at the time of writing (see de Souza, White, & Springer, 2016; Springer, 2014, 2016; Springer, White, & de Souza, 2016; White, 2015b; White & Springer, 2018; White, Springer, & de Souza, 2016).

## *Session 1: Thinking about Critical Animal Geographies and Morality*

The opening session explored the "moral turn" in human geography, and explicitly focused attention toward the ways in which nonhuman animals are (ab)used in Western society. An explicit open discussion on the more violent and exploitative relationships came to the fore as the session developed. At the initial stage, I wanted to create an open and "safe" space within the classroom, in order to encourage students to engage and actively participate, eagerly and willingly. This helped not only to encourage deeper, richer, and more nuanced discussions in the class, but it also meant that many of the key arguments and themes explored were also "owned" collectively by the group present. One approach that I've found particularly successful here has been to stand at the front of the class, armed with whiteboard marker (or chalk), and invite students to think about: (a) the nonhuman animals that are personally known in some way to them, and to think about the spaces that they would find these animals in; (b) other spaces that nonhuman animals are found; what these spaces are called, and to think about the lived experiences of the animals within these spaces.

Having given the class a few minutes to reflect on these opening questions, I invited comments and responses, and made these visible and tangible by highlighting them on the whiteboard. The opening response was typically "pets." Note, here and elsewhere I do take the time to critique any problematic reference points (e.g., "pet," "farm animal," "livestock") to encourage students to recognize the negative signifiers that these embody. So "pet/companion animals" are scribbled on the whiteboard, with hashed lines drawn around the word. Where are "pets" found? "The home!" What kind of animals do we consider to be "pets?" Typically, here the spectrums of animals range from "domestic" (dogs, cats, rabbits, guinea-pigs, hamsters, budgies, goldfish) to more "exotic" species (snakes, terrapins, tarantulas, etc.). Students reflect on the positive emotional relationships that are created through contact with these nonhuman animals—"they are part of the family," "I love them," "they are irreplaceable," "I was devastated when my sick cat died." I ask a rhetorical question: "Do you name these animals ?" Of course!

The positivity and general willingness to engage in the session in these early stages is palpable. The subject material is both unexpected—"*I didn't think I'd ever be talking about my sick guinea pigs in a human geography lecture*"—and fosters a common and often excitable dialogue within the class. Moving away from "the home" other examples of animals and the spaces and places they are found are then noted on the board. The five main nonhuman animal-spatial groupings include:

- *Animals at schools.* Fish, gerbil, hamster. Students again laugh and talk about "their" schools' animals—some might reflect fondly on having the responsibility for looking after them at home on a weekend or school holiday. There was a shared consensus that these animals (here and in the home) impressed upon humans the need to "care" and be compassionate: to help and protect those in need of help (by being kind, providing food, water), and so on.
- *"Farm" animals.* Cows, pigs, sheep, chickens, turkeys. Students reflected on "educational" trips to farms at primary school (i.e., before the age of 9 years). Most had never visited a farm since. None had ever seen the inside of a "factory farm," or slaughterhouse.
- *"Vivisection."* Reference to hidden spaces: private spaces. A range of animals were identified: rats, mice, but others less known (cats, dogs, horses, pigs, etc.). The latter caused some anxiety in the class—"*I didn't realise there were so many animals still experimented on, and within universities: it's horrible!*"
- *"Caged animals."* Most students had visited a zoo, either as part of a school trip or with family. Mixed reaction—some hated seeing the way in which certain animals (especially lions) were kept. Others found it a fun and worthwhile experience.
- *Animals as "entertainment":* Including circus animals, horse-racing, and blood-sports, fishing, fox-hunting. Some students participated in these actively. Other European examples of animals were highlighted here, with particularly Spain and bull-fighting.

The spaces were outlined with deliberately hatched lines, allowing connecting overlaps to be made between "these animals" and "these spaces"; highlighting contrasting animal conditions and relationships that are fostered here; demonstrating the relative proximity of these spaces to each other (public, private). Their visual depiction on the crowded whiteboard made an impression on the students. Students commented on the complexity of these relations, noting the ways in which space functioned to encourage attachment and bonding relations, to those that fostered the opposite. Remarks were also forthcoming about how certain species of animals rarely moved between these spaces: for example, we have a clear impression of what type of species "pets" are, what "farm" animals are, and so on.

In attempting to problematize complex sociospatial relationships with other species, the class were then invited to think about this from an ethical or moral perspective (see Pedersen, 2004). Is there anything morally "wrong" with the ways animals are treated by humans in the examples given? Here the students were encouraged to reflect on Mark Rowlands (2002) very accessible and important book: *Animal Like Us*, by thinking about morality as a club. The key reasoning here is that "we view [all humans and] many animals as things that are *morally considerable*" (2002, p. 26). However, Rowlands takes issue with the second-class status that many/all nonhuman animals are assigned within this club, a stance which he finds is morally unacceptable position:

> It is morally unacceptable because it is *inconsistent*. Inconsistent with what? Basically, with the rules of the moral club. If we checked the club rules, we would see that a radical egalitarianism holds within the club. All members are equal members. And so it is unacceptable, morally unacceptable, to accord one member less consideration than another. (2002, p. 28)

To focus attention—and inject a deal of urgency into proceedings—a small group-based discussion around that powerful pedagogical device so loved by philosophers: *The Lifeboat Scenario* was introduced. Divided into small groups of four or five, the students were invited to consider the following Lifeboat scenario (Table 7.1) and: (a) respond individually to the guidance and instruction given, and then (b) try to come to some consensus within their groups as to which five persons they would save. They were given 15–20 minutes.

*Table 7.1:* The Lifeboat Scenario (2016 version). Source: Author.

| | |
|---|---|
| Imagine you are onboard a large ship that is rapidly sinking at sea. There is just one lifeboat left and you have to decide the order in which the following persons should be given priority on the lifeboat. Who would/should you save first? Who would/should you leave behind? What moral criterion/criteria can you give for justifying prioritizing one individual (human or nonhuman) over another? *You are asked to consider the following when deciding who to include on your lifeboat/leave behind on the ship. You know only the following information:* | |
| 1 | **Charlie** is a young, friendly Labrador. He is the guide dog of a young teenage girl called Sally. |
| 2 | **Simone** is a 26-year-old white cis-female. She is a famous tennis player, married, and is the mother to a six-month-old daughter. She does a lot of work for charity. |

| | |
|---|---|
| 3 | **Hannah** is a well-behaved 3-year-old human toddler, who loves painting. This is the first time her parents have taken her onboard a ship. (*It is not known where her parents are at this time.*) |
| 4 | **Babe** is a young male pig. He has been the subject of recent media attention, owing to the fact that he has demonstrated an intelligence that p*uts him at least on a par* with a four-year-old child, and considerably higher than many other nonhuman animals (including dogs and cats). |
| 5 | **Alice** is an 85-year-old African-American woman. She has been saving up for many years now to afford to the trip overseas to see her great-grandchildren for the first time. |
| 6 | **Tommy** is a 5-month-old baby, who gurgles a lot. His mother is currently safe on another lifeboat, but frantic with worry. It is not known where his father is. |
| 7 | **Paul** is 39. He is a highly intelligent and a healthy human. Paul has only been out of prison for a year, having been incarcerated for 20 years for the murder of a 15-year-old youth. |
| 8 | **John** is 53 years old—he has been in a coma for the past two years—and he will almost certainly never come out of this coma. |
| 9 | **Bob** is a bachelor, aged 40, who has invested millions into a Sheffield football team. He is looking to bankroll a new regeneration initiative which will work with helping disadvantaged youths from a deprived suburb of Sheffield. |
| 10 | **Samantha** is a 22-year-old pop phenomenon. She is currently the No.1 best-selling artist in the UK and USA. |
| 11 | **Frank** is an 8-year-old boy, and has severe mental health impairments. |
| 12 | **Ryan**, a heavy smoker and drinker, is an 18-year-old white male. He is a keen and outspoken member of the British National Party (BNP). |
| 13 | **Louisa** is a 20-year-old geography student at a prestigious northern UK university (Sheffield Hallam). She is ambitious and fiercely committed to her studies and would like to do some overseas volunteering when she graduates. |
| 14 | **Adrian** is an attractive 19-year-old history student at a prestigious southern UK university. He doesn't really care whether he passes his course or not, as he would much rather spend his youth (and money) on drink and partying. |
| 15 | **BoBo** is one of only two remaining Rwandan mountain gorillas in the world. She is being taken to the other gorilla, in the hope that the pair will breed and save the species from extinction. |

For many students, this innovative and unusual scenario quickly took on a serious quality. Effectively, many students found the responsibility to make "the right" decision in this pressurized situation incredibly challenging.

Though the majority of students engaged fully, some students were palpably frustrated and angry at what they were being asked to do. For example, one student spoke up in class and said: "*I'm not doing this, you're asking me to play God, and I'm not doing it.*" Another said, "*It's just arbitrary: you may as well toss a coin to decide.*" For many, though, responding in the spirit of this critical exercise was provoking, stimulating, and ultimately came with a real sense of accomplishment.

When we came to debate the arguments and counterarguments collectively as a class, unsurprisingly, the discussion was rather intense, and opinions were polarized and often heated. Unsparingly, the question of whether the nonhuman animals: BoBo the gorilla, Babe the pig, and/or Charlie dog should be saved was particularly contested. Most students over the years have instinctively left the nonhuman animals behind, citing the "moral" case that "*they're not human*!" to justify their reasoning here. Questions of quality of life, though, come into play, and indeed do politics. Ryan is frequently condemned to be left behind because of his right-wing sensibilities, and Paul, because of his past. All of this, needless to say, is rich material for the critical animal pedagogies engaged in this session, and for future reflection.

To problematize the appeal to "species," the students were encouraged to reflect on the fact that this constitutes an appeal to biological qualities. They were asked if they were comfortable with this, or if this is to be held up as a moral category, then what more uncomfortable scenarios may emerge, given that sex and race are both biological categories. If they are proposing species as a moral relevant criterion, then how would they argue against somebody who approaches this with a racist and/or sexist worldview: the outcomes of which would be morally abhorrent to them? This unsettling thought, and the implications that emerge from this exercise and conversations, are deliberately kept open for further reflection (and through further reading). There were no final conclusions, or particular suggestions as to what the student should be thinking about. The only other point impressed upon the students is that these hermeneutic devices should not be seen as directly transposable to "real"-life praxis: As Rowlands reminds us:

> (M)ost of the time the [moral] decisions we have to reach about our treatment of animals *are not even close*. There is no need to agonize over the precise balancing of potential benefits versus potential losses. *It's not even close*. Unlike lifeboat scenarios, it's not as if there are vital human interests at stake which we have to weigh carefully against the similarly vital interests of animals. Most of the time, our mistreatment of animals stems from our letting relatively trivial human interests outweigh even the most vital interest of animals. Nowhere is this more so than in our raising and killing of animals for food. (2002, p. 99)

Broader reading and reflection between the sessions are encouraged, and from focusing on critical human–animal geographies, in order to impress upon the importance of capturing the entanglement of interspecies relations vis-à-vis violence, exploitation, and domination, the module moves to focus on questions of intersectionality.

## *Session 2: Intersectionality*

> I remember at the time feeling like this was a really new side to geography, a pioneering side, a very contemporary side, and anything that makes the audience feel uncomfortable and reflective is an exciting part to any field of study... At the time I was one of the only vegetarians I knew, I think I knew around three! I had been vegan in the year before, but, without a single vegan friend and always bearing the harassment of having to answer questions from defensive meat eaters I gave it up, but for the two years before I'd really become an incredibly vocal and passionate feminist, and I was so very proud to discuss it, I was really bristling for a fight because, even at university, I never really found any positive feminist connections! So Richard's Philosophy, Space and Place lectures really excited me, and when I got the reading list I bought quite a few of the books, including "The Sexual Politics of Meat" which I remember reading in just three days or so. (Natasha, undergraduate student, 2011)

Having problematized the moral geographies emerging from the dominant ways in which we think about, and use, the following session, focus on the intersectional nature of violence and oppression. The minimum learning outcome here is to understand how animal liberation is human liberation (and vice versa), that the two are deeply connected and intertwined on so many levels. The particularly important influence of feminism and eco-feminist thought has been central in defining the grounds and trajectories here (Cudworth, 2015; Kemmerer, 2011). There are many different readings that the students draw on to articulate the intersectionality between humans and nonhuman animals, but this passage from Alice Walker (1988, pp. 7–8) never fails to stir some emotional and visceral response:

> I almost laughed (I felt too sad to cry) to think there are people who do not know that animals suffer. People like me who have forgotten, and daily forget, all that animals try to tell us. "Everything you do to us will happen to you; we are your teachers, as you are ours. *We are one lesson*" is essentially it, I think.
>
> There are those who never once have even considered animals' rights: those who have been taught that animals actually want to be used and abused by us, as small children "love" to be frightened, or "women love" to be mutilated and raped.... They are the great grandchildren of those who honestly thought, because someone taught them this:

> Women can't think, "and niggers can't faint."
>
> ...Once a friend came to visit and said, looking out on the soothing view: "And it *would* have to be a *white* horse; the very image of freedom" And I thought, yes, the animals are forced to become for us merely "images" of what they once so beautifully expressed. And we are used to drinking milk from containers showing "contented" cows, whose real lives we want to hear nothing about, eating eggs and drumsticks from "happy" hens, and munching hamburgers advertised by bulls of integrity who seem to command their fate. As we talked of freedom and justice one day for all, we sat down to steaks. I am eating misery, 1 thought, as I took the first bite. And spit it out.

Drawing closer attention toward intersectionality, and the intersections of "sexism" and "speciesism" in particular, the sessions here worked with many of the core themes and arguments that Carol Adams (2002) focuses on in *The Sexual Politics of Meat*. A similar pedagogic approach to the first session was adopted, aiming to capture the broad experiences, understandings, perceptions, and thoughts of the class at the beginning of the session. One of the most effective and disarming ways to do this was to problematize the everyday use—and power—of language. For example, I invited the class to think about how women are "animalized" through reference to nonhuman animals. There is often self-conscious, awkward laughter at the beginning. Examples are slowly volunteered: "Fat Cow! Ugly Pig!" and it has often been necessary to remind the class that this is a "safe-space," insofar that they can be encouraged to be open and honest without fear of recrimination or consequence. With the example of "the pig" or indeed "the cow," this is another opportunity to highlight what society (our cultural norms) has encouraged us to think of this animal. Almost all are negative: in addition to "fat," "stupid" (*pig ignorant*), "untidy" (being told as a child that your untidy bedroom is "*like a pig sty*"), and so on. Moreover, in the last two years, I've asked for all animal referents of the human body to be made. The often-positive sexual connotations of the male body: "hung like a donkey," "trouser snake," "wild bull," "stallion"...as well as sexual acts "at it like bunnies", "doing it doggy-style" serve to emphasize these uneven gender-species divisions more profoundly. Observing the students there is an uncomfortable and heady combination of laughter, mixed with genuine realization of the serious point being made. These examples may instinctively be seen as both funny and not funny, but when they are taken as more-than-a-sum-of-their-parts, the profound significance and power of language, in how women and nonhuman animals are categorized and demeaned, quickly take hold.

Faced by this unsettling juxtaposition of mainstream language and imagery *in this context* is something that most students find both deeply unsettling and revelatory any speciesist human-reflex defense "they're only animals" seems particularly asinine. Being encouraged to bring a critical reading to these adverts, and the way in which they explicitly draw on a human-centric and male-gaze to achieve this, makes a deep and lasting impression. Rebecca (undergraduate student, 2014) provided one example of this:

> The struggle facing women and animals in terms of their place in life/society registered more clearly as a result of the lectures—previously I hadn't made such an obvious connection.... Even being a vegetarian still gives me a conscience about supporting such patriarchal, unethical and unquestioned system(s)—makes me think should I be a vegan?

One of the key debates I've wrestled with has been the use of video images depicting extreme human-on-animal violence, even more so than the human-on-human violence that I show elsewhere (e.g., of police brutality, or of state-sanctioned acts of terror). Quite why this should be the case, I'm not so sure. Perhaps, it is because regarding nonhuman animals there is far more complicity and support in the classroom with the abuse of nonhuman animals (killed for food, for example). The decision has been to highlight short clips, some of which draw attention to explicit acts of violence (e.g., routine abuse of nonhuman animals on so-called "organic" farms) to incredibly powerful clips where the shadow of violence lingers, but is never seen. Watching the faces of the class react to these videos has been an experience. Some close their eyes completely, try to look away, or look down. Others have a look of horror, sit silently and attentively.

> At the time I had just read the book "Wild" by Jay Griffiths and I recall seeing a similarity between the insult of "being wild" and "they were acting like a girl", and "like a wild animal", there was this idea that women are uncontrollable and emotional. But I was proud to be emotional, to be passionate, and if this is what the insults suggest—to be uncontrollable. Richard's lectures drew capitalism into the equation, that patriarchy benefits from selling women and meat as similarly abusable because it bolsters masculine identity, and certainly the visual evidence (advertisements mostly) was very moving and convincing, and sickening. (Natasha)

A significant thread that entangles both human and nonhuman animals in a web of suffering and violence is highlighted: *capitalism*. Here the voices and experiences of human animals working in slaughterhouses are brought center stage, again in ways that enforced the interspecies geographies of violence that the class found deeply disturbing and provoking.

### *Session 3: Vegan Praxis Activism and Politics of Total Liberation*

The third and final session is designed to be as constructive, open, inclusive, and empowering as possible. The overriding message here was the need to try to better recognize, and take seriously, the complex links and intersections between all forms of struggle. Certainly, it was impressed upon the Human (sic) Geography students focusing critically on the relationships with other animals not discrete or separate from the many "human struggles." As Ryder (2001) argued: "The struggle against speciesism is not a side-show; it is one of the main arenas of moral and psychological change in the world today. It is part of a new and enlarged vision of peace and happiness" (p. 1).

Speaking about animal rights activism, I reflected on my own experiences and privileges, as well as my personal journey from (behavioral) omnivore to vegetarian (at university) and vegan. From eating nonhuman animals to *not* eating them, not wearing their skins, nor consuming their products (milk, eggs). I try to empower students in a way that takes to heart the thoughts of Andrzekewski (2003, p. 24):

> I reminded them that we are here to learn and to decide for ourselves what actions we want to take to work for a peaceful, nonviolent world. We cannot force others to make these changes, but we can encourage them in many supportive ways to do so. I suggested that we consider ourselves on a journey, rather than a deadline; that there is much to study and understand, and thousands of ways we can change if we are really committed to eliminating the suffering and exploitation of animals. This discussion helped to calm people's emotions and helped them reflect upon how they would want to be treated themselves.

For those who argue that they cannot make a difference given the scale of (abuse) involved, I respond with Rowlands' (2002, pp. 179–180) argument, "It is always in our power to *not* do something. If our eating animals contributes to a husbandry system that inflicts unnecessary physical and psychological suffering on millions of animals, and it is always within our power to not eat animals, then that is what we should do: not eat them."

More than make (relatively easy) lifestyle changes, other important ways in which students can take direct action were explored. Several students expressed an interest in finding out more about local "animal" activist groups, including Sheffield Animal Friends (whose motto is *educate, agitate, liberate*!), and local hunt saboteur groups. There are many ways though that these ideas take root, and establish themselves, sometimes in unexpected ways. Of these, arguably the most satisfying, encouraging, and important outcomes of the conscious-raising that takes place in the *PSP* lectures is embodied in how students then engage in intersectional forms of activism within and beyond the academy. Within their future academic studies, for example, students taking the module have gone on to further deepen their engagement with critical animal geographies, and critical animal studies more generally, in their final year 10,000 word dissertation. This connection is illustrated in the reflection below:

> The emotionally challenging and shocking nature of some of the topics covered were one of the unique things about this module, allowing me to see things on a personal level. The sessions given on "women as meat" and the use of derogatory advertising campaigns as a marketing technique for me were some of the most memorable. …
>
> This sparked a personal interest to focus on other messages that could be seen in everyday circumstances that carried the same derogatory themes towards women. This led me to carry out my dissertation research project on a related tangent to this and look at the disparaging displays and treatment of women, often linking back to the relation of "women as meat" in nightlife spaces and club marketing posters. I found being able to study such a contemporary subject that is constantly changing to be both emotionally challenging but also hugely rewarding. (Ellie, undergraduate student, 2015)

Many of these dissertations have been exemplary in their academic standard, and originality. Not only have these attracted some of the highest grades ever awarded, but they also have been put forward for various university and national undergraduate research prizes.

Though there is a silent majority that dominates the student groups that have engaged with the module, insofar as animal rights activism "beyond the academy" is concerned, there are notable exceptions. Several students, recognizing the inherently abusive nature of the dairy industry, have been moved to rethink, and reject their vegetarianism, and embrace a vegan diet. Others have moved from a meat-and-dairy diet to either vegetarian (with the intent of "going vegan") or have embraced a vegan diet from the get-go. Encouragingly many students have recognized the importance of not reducing veganism to mere "food choices," rather to approach veganism as part of a broader activist praxis, and a true counterstrategy to capitalism (see Dominick, 2015). Interest in joining local animal rights groups (on campus and in the city) has been seen, as have campaigns to increase more vegan options in the university's cafes and restaurants. In this context, however, what has caused something of a key dilemma has been the absence of direct support, perhaps "aftercare," for students that have embarked on this life-changing, life-affirming, perhaps, journey. Two students come quickly to mind here. Having made the decision to go vegan, both students were subjected to a variety of "easy-going," "only joking!" bullying from friends and family, of a similar nature to that reflected on in the opening of this chapter. The social unhappiness that the students had to endure was appalling. One continues to be vegan, noting that the rise of social media groups (e.g., Sheffield Vegan Society on Facebook) was invaluable in giving them access to the strength, support, and solidarity that she so dearly needed. The other regrettably came to the conclusion that they couldn't maintain this commitment to nonviolence When eating alone he would not buy animal flesh, but in a group situation (university friends or family) would "just eat whatever…it's not what I want, but for the moment at least it's what I'll do."

## *Conclusion/Reflections*

> Although I've had a long-standing passion for animal rights and other social justice causes, more broadly, it wasn't until I came to university and had the opportunity to take the "Philosophy, Space, and Place" module, which includes sessions on "Critical Human-Animal Geographies" that I began thinking and writing about these issues in a critical, scholarly and interdisciplinary way; with a teacher and in an environment which was wholly encouraging about the topic. I remember being excited that someone was giving the topic the serious consideration it deserves (and requires), something which had since been absent throughout my educational experience so far. (Lindsey, undergraduate student, 2008)

> I think this is what this module really achieved in many of the students: having to explore a very different perspective, but one which is so relevant to everything we see, eat, and experience every day. Our "humanity," consumption, and gender run underneath everything we do, yet many of us had never considered the ideas put forward in this series of lectures. (Natasha)

David Pellow (2014, p. xv) referred to academia as a "Space of Liberation and Repression." Despite the enormous challenges that the university poses in terms of conscious raising for an intersectional social justice approaches for liberation, there are many important and rewarding and empowering opportunities that also present themselves. Speaking to this, the chapter has provided just a very brief overview of some of the central experimental critical pedagogies and approaches found within one undergraduate module. Given the student-centered forms of learning here, it is important to state explicitly that no one session is ever the same: you try and work with whatever constellation of perceptions, experiences, and values are present in that pedagogical space. Embracing the unknown, being open to teaching (and being taught!) *in the moment* is absolutely crucial to be able to engage with the spectrum of attitudes and expectations that student have toward other animals. It is difficult and exhausting work, but in many ways this is as it should and must be. The hope is that the seeds of these liberatory futures, which carry with them the promise of love, empathy, beauty and nonviolence, will find fertile grounds in people's hearts and minds. And, having taken root, will flourish in the ongoing struggle(s) for social and spatial justice.

## References


Adams, C. J. (2002). *The sexual politics of meat: A feminist-vegetarian critical theory* (10th ed.). New York, NY: Bloomsbury.

Andrzekewski, J. (2003). Teaching animal rights at the university: philosophy and practice. *Journal for Critical Animal Studies, 1*(1), 17–35.

Cudworth, E. (2015). Intersectionality, species and social domination. In A. J. Nocella II, R. J. White, & E. Cudworth (Eds.), *Anarchism and animal liberation: Essays on complementary elements of total liberation* (pp. 93–107). Jefferson, NC: McFarland & Company.

de Souza, M. L., White, R. J., & Springer, S. (Eds.). (2016). *Theories of resistance: Anarchism, geography and the spirit of revolt.* Lanham, MD: Rowman & Littlefield.

Dominick, B. (2015). Anarchy-veganism revisited. In A. J. Nocella II, R. J. White, & E. Cudworth (Eds.), *Anarchism and animal liberation: Essays on complementary elements of total liberation* (pp. 23–39). Jefferson, NC: McFarland & Company.

Drew, L., & Socha, K. (2015). Anarchy for educational praxis in the animal liberation movement in an era of capitalist triumphalism. In A. J. Nocella II, R. J. White, & E. Cudworth (Eds.), *Anarchism and animal liberation: Essays on complementary elements of total liberation* (pp. 163–178). Jefferson, NC: McFarland & Company.

Kahn, R. (2003). Towards Ecopedagogy: Weaving a Broad-based Pedagogy of Liberation for Animals, Nature, and the Oppressed People of the Earth. *Animal Liberation Philosophy and Policy Journal, 1*(1), 35–52.

Kemmerer, L. (2011). Theorizing 'others'. In N. Taylor & T. Signal (Eds.) *Theorizing Animals: Re-thinking Humanimal Relations* (pp. 59–84). Leiden, Netherlands: Brill.

Lupinacci, J. (2015). Recognizing human supremacy. In A. J. Nocella II, R. J. White, & E. Cudworth (Eds.), *Anarchism and animal liberation: Essays on complementary elements of total liberation* (pp. 179–193). Jefferson, NC: McFarland & Company.

Nocella, A., Sorenson, J. Socha, K., & Matsuoka, A. (2014). *Defining critical animal studies: An intersectional social justice approach for liberation.* New York, NY: Peter Lang.

Nocella, A., White, R. J., & Cudworth, E. (2015). *Anarchism and animal liberation: essays on complementary elements of total liberation.* Jefferson, NC: McFarland & Company.

Pedersen, H. (2004). Schools, speciesism, and hidden curricula: The role of critical pedagogy for humane education futures. *Journal of Future Studies, 8*(4), 1–14.

Pellow, D. N. (2014). *Total liberation: The power and promise of animal rights and the radical earth movement.* Minneapolis, MN: University of Minnesota Press.

Rowlands, M. (2002). *Animals like us*. London, UK: Verso.

Ryder, R. D. (2001). *Animal revolution: Changing attitudes towards speciesism*. London, UK: Bloomsbury Academic.

Sorenson, J. (2011). Constructing extremists, rejecting compassion: Ideological attacks on animal advocacy from right and left. In J. Sanbonmatsu (Ed.), *Critical theory and animal liberation* (pp. 219–238). New York, NY: Rowman and Littlefield.

Springer, S. (2014). Why a radical geography must be anarchist. *Dialogues in Human Geography*, *4*(3), 249–270.

Springer, S. (2016). *The anarchist roots of geography: Towards spatial emancipation*. Minneapolis, MN: University of Minnesota Press.

Springer, S., White R. J., & de Souza, M. L. (2016). *The radicalization of pedagogy: Anarchism, geography and the spirit of revolt*. Lanham, MD: Rowman & Littlefield.

Walker, A. (1988). Am I blue. In A. Walker (Ed.), *Living by the word: selected writings 1973–1987 by Alice Walker*. San Diego, CA: Harcourt Brace Publishers.

White, R. J. (2015a). Following in the footsteps of Élisée Reclus: Disturbing places of species violence that are hidden in plain sight. In A. J. Nocella II, R. J. White, & E. Cudworth (Eds.), *Anarchism and animal liberation: Essays on complementary elements of total liberation* (pp. 212–230). Jefferson, NC: McFarland & Company.

White, R. J. (2015b). Animal geographies, anarchist praxis, and critical animal studies. In K. Gillespie & R. C. Collard (Eds.), *Critical animal geographies: Politics, intersections, and hierarchies in a multispecies world* (pp. 19–35). New York, NY: Routledge.

White, R. J., & Springer, S. (2018). Making space for anarchist geographies in critical animal studies. In J. Sorenson & A. Matsuoka (Eds.), *Critical animal studies*. Lanham, MD: Rowman & Littlefield.

White, R. J., Springer, S., & de Souza, M. L. (2016). Performing anarchism, practicing freedom, pursuing revolt. In R. J. White, S. Springer, & de Souza, M. L. (Eds.), *The practice of freedom: Anarchism, geography and the spirit of revolt* (pp. 1–22). Lanham, MD: Rowman & Littlefield.

# 8. *Teaching the Animal in Foreign Languages: An Ecopedagogical Approach*

ANITA DE MELO

In the 21st century, there has been much discussion on how education should focus on a learning process that promotes immediate change to our societies, a learning process that helps students become critically aware of world issues and participate actively in resolving them; this approach is often referred to as Global Education (Maastricht Declaration on Global Education in Europe, 2003). Global Education advocates for an education that addresses global justice, human rights, sustainability, peace, and intercultural communication, and it must be based on tolerance, solidarity, equality, justice, inclusion, cooperation, and nonviolence (Global Education Network, 2009).

The Teaching of Foreign Languages (TFL) today also transcends target language proficiency and focuses on a teaching approach similar to that advocated by Global Education. There has been continuing debate and projects that support Global Education in TFL, based on an interdisciplinary methodology to language learning that brings students into contact with communities speaking the target language. Such encounters encounters would facilitate cultural and linguistic comparisons, would promote positive language proficiency, and would also prepare students to become global citizens, engaged in resolving global issues (Della Chiesa, Scott, & Hilton, 2012; Jacobs, 2010; Yakovchuk, 2004). In this chapter, I argue that Global Education in TFL can and must expand its range of social concerns to include nonhuman animals' issues.

From the second half of the 20th century the orientation of the Humanities has been to question this once taken-for-granted anthropocentrism (Guha,

2000; Plumwood, 1996). And the study of nonhuman life issues, known also as environmentalism, is becoming increasingly a crucial subject in the disciplines of History, Anthropology, Philosophy, Literature, Sociology, and Cultural Studies (Berkhout, Leach, & Scoones, 2003). This new environmentalist emphasis shaping the Humanities must not be incompatible to the current and extensive issue of animal exploitation.

At the present time, and at an unprecedented rate, humans heavily depend on nonhuman animals to satisfy their needs, and this exploitation causes violence, a staggering amount of suffering, and also destruction to the environment. According to the Australian philanthropist Phillip Wollen (2012), what humans do to animals is the greatest social injustice of modern times. Although achieving a fair and just relation between society and the animal might seem visionary and almost impractical due to our millenary dependency on animals along with our speciesist attitude toward nonhuman life along with our political and economic capitalist systems of exploitation, the extensive gap between ethics and the treatment of animals requires a fundamental and revolutionary social change, one in which Global Education must have a central role a central role. Hence, it is my assertion that Global Education in TFL must make the topic of animal exploitation a curriculum priority. Such a shift in focus critically challenges the normalization of animal exploitation and creates an atmosphere that awakens students' empathy for animals. It is hoped that this will also create opportunities for social change vis-à-vis animal maltreatment. Source materials for my arguments are drawn from Paulo Freire's ideas of Critical Pedagogy (1972); the international ecopedagogy movement as theorized by Richard Kahn (2010); and from the articles on animal liberation edited by Steven Best and Anthony J. Nocella II (2004).

In his groundbreaking work *Pedagogy of the Oppressed*, Paulo Freire (1972) argues about the necessity of an emancipatory dialogic pedagogy of the oppressed, in opposition to the pedagogy of the ruling and dominant classes; a pedagogy that contributes to the liberation and transformation of those socially oppressed; such liberation would, consequently, transform society (p. 3). Based on his own experience working with adult literacy in Brazil, the oppressor Freire is referring to is the classes of people who are denied access to the "word" (formal education) and who are, then, forced to see the world through the lens of the oppressor, the dominant and ruling classes who "name the world" (p. 32). Freire argues that it is paramount that this unjust cycle between oppressor and oppressed be broken, and be achieved through "two distinct stages. The first, the oppressed unveils the world of oppression and through praxis commit themselves to its transformation" (p. 80). In other words, Freire claims that the oppressed must have access not only to a formal education but also to a revolutionary one. Frequently, institutionalized models of education and teaching, which Freire calls the "banking concept" (p. 21), pass on the oppressor ideology. A critical pedagogy of teaching must be constructed upon dialogue, and dialogue, Freire argues, is the "the encounter between men, meditated by the word, in order to name the world" (p. 32). However, the school systems, Freire insists, collaborates with the *status quo* in which the dominant and ruling classes name the world on behalf of the socially oppressed.

Freire does not discuss in depth the idea that oppression and layers of power are firstly perpetrated through one of the basic levels of our collective identity, that is, our social and historically constructed languages. About a decade later, after the publication of his *Pedagogy of the Oppressed*, Freire would engage in dialogues with Donaldo Macedo about the role of subjectivity in an emancipatory literacy. Macedo was concerned that the individual consciousness might be "at odds with the collective consciousness" (Darder, Baltodano, & Torres, 2003, p. 354). Freire elaborates that "it is impossible to deny the constitutive power of consciousness in the social practice" (p. 355). Freire, then, points out that subjectivity is constructed from the objective world one lives in. Language is socially constructed throughout time, and one is born into it. Freire warns against a "mechanical comprehension of language" (p. 358). He provides a Marxist perspective suggesting that "one cannot understand and analyze a language without a class analysis (p. 358), for the reinvention of power is attained by the reinvention of production, one in which all segments of society would participate through a reinvention of culture and *language* [my italics]" (p. 360). Our maltreatment of nonhuman animals has its origins, at a basic level, throughout our human languages, which are transmitted intergenerationally. And this is precisely why courses on languages concerned with global issues must adapt a critical pedagogy, instigate dialogue, and critically challenge, through a counterdiscourse, the hegemonic institutionalized language of dominance (in the process of naming the world) over animals. Critical pedagogy and Global Education can facilitate the TFL into a powerful tool in presenting animal issues.

Although *Pedagogy of the Oppressed* is one of the most important books of the 20th century, the concept of problem-posing in education (Freire, 1972, p. 26) proposed by Freire is still, in many instances, an aspiration. It is important to note that educators too suffer from our institutionalized systems of education. In *Life in Schools* (1989), Peter McLaren formulates an extensive theoretical introduction to critical pedagogy. This work is a late edition of McLaren's diaries published in 1980 as *Cries from the Corridor* in which the author recounts his teaching experiences in an elementary school on the outskirts of Toronto. *Life in Schools*, McLaren states, is "the story of his reinvention as an educator, from a liberal humanist who focused on social and educational reform to a Marxist humanist advocating revolutionary praxis, that is, educating for social revolution through critical pedagogy" (Ross, 2016). In North America McLaren's work is pioneering and revolutionary. Like Paulo Freire, McLaren worked with the oppressed, both the children who lived in the housing projects in Toronto and the schoolteachers who had to deal with overcrowded and unfunded classrooms (Ross, 2016). By revisiting his early teaching approaches and by developing a theoretical analysis of critical pedagogy later on, McLaren's *Life in Schools* "prompts readers to begin to analyse the assumptions that inform their teaching practices and, more generally, the liberal humanist discourse of contemporary progressive and well-intentioned teachers" (Ross, 2016). McLaren's dialectical approach to investigating the school system is essential for those interested in Global Education as a way of fighting global social injustices. "Viewing schools as democratic public spheres means regarding schools as sites dedicated to forms of self and social empowerment..." (McLaren, 1989, p. 238). Echoing Freire's idea that there is no neutrality in education, McLaren warns that "abandoning the project of political engagement and human possibility, educators and professional critics alike have promoted despair at the expense of analysing the fundamental relationships of power that characterize existing forms of cultural and political hegemony" (p. 240). In other words, "the teacher's task must take the form of a critical pedagogy" (p. 241), that is, as educators we must take the side of the oppressed, although this is a "perilous and immense task" (p. 242).

It can be argued that Global Education is one aspect of this difficult and, at times, vulnerable work of critical pedagogy. As in critical pedagogy social injustices must first be uncovered and understood, and then through praxis comes liberation which is "the action and reflection of men and women upon their world in order to *transform* it" (Freire, 1972, p. 26). This pedagogy, Freire argues, has two specific stages:

> In the first the oppressed unveil the world of oppression and through the praxis commit themselves to its transformation. In the second stage, in which the reality of oppression has already been transformed, this pedagogy ceases to belong to the oppressed and becomes a pedagogy of all people in the process of permanent liberation. In both stages, it is always through action in depth that the culture of domination is culturally confronted. (p. 8)

Specialists in TFL began to work with Freire's idea by the end of the 1970s (Crawford, 1978). Today critical pedagogy in TFL is an important part of language curriculum theory and instructional practices that support and advance teaching and the study of languages in ways that promote social justice (Crookes, 2012). TFL is viewed not only as a new instrument to express or communicate; it becomes a product constructed by how language learners can "recognize themselves, their social surroundings, their histories, and their potentialities for the future" (Riasiti & Mollaei, 2012, p. 223). Global Education in TFL has incorporated successfully a variety of methodologies with the aim of preparing students to become global citizens. Some of these methodologies include Service Learning (Weldon & Trautmann, 2003), sociolinguistic tools delivered to heritage students (Correa, 2011) with interdisciplinary and cross-curriculum collaboration (Magnin, 2002). Additionally, many colleges and universities that teach foreign languages, in the United States at least, offer some type of study abroad programs (SAPs) in which students can experience full cultural immersion. Graduate students in upper level courses of foreign language majors have greater experience in examining social issues through cultural and literary readings.

## *The Ecopedagogy Movement*

The 2005 UN-Millennium Ecosystem Assessment (MEA) released a study on the planet's ecology. The study states that over the span of the past fifty years, humanity has dramatically changed Earth's ecosystems, causing unprecedented destruction of world forests and causing the extinction of many species of animals. "Between 10% and 30% of mammal, bird, and amphibian species are currently threatened with extinction" (MEA, 2005, p. 18). The assessment explains that a major reason for the extinction of species is habitat loss, and even if "habitat loss were to end today, it would take hundreds of years for species numbers to reach a new and lower equilibrium due to the habitat changes that have taken place in the last centuries" (p. 102). Among the many reasons for habitat loss, MEA points out that the "[E]xpansion of livestock production around the world has often led to overgrazing and dryland degradation, rangeland fragmentation, loss of wildlife habitat, dust formation, bush encroachment, deforestation, nutrient overload through disposal of manure, and greenhouse gas emissions" (p. 610). Human-caused changes in the Earth's ecosystems affect directly the quality of human life, contributes to climate change, and the extinction of species. Living in a sustainable world must be a global concern, a priority in which education must play an important role.

The ecopedagogy movement, as theorized by Richard Kahn (2010), derives from Paulo Freire's revolutionary ideals of critical pedagogy, that is, it is committed to the firm belief that education must help students to think critically and make connections between their own individual problems and the social context in which they are immersed in order to achieve total liberation. The ecopedagogy movement also "seeks to interpolate quintessentially Freirian aims of the humanization of experience and the achievement of a just and free world with a future-oriented ecological politics...grounded in normative concepts such as sustainability, planetary, and biophilia" (p. 18). Although ecopedagogy also has its origins in the global south, the discussion originated from discussions during the Rio Earth Summit of 1992, and later it would "additionally incorporate more typically northern ecological ideas such as the intrinsic value of all species, the need to care for and live in harmony with the planet, as well as the emancipatory potential contained in human aesthetic experiences of nature" (p. 19). Thus, the ecopedagogy movement recognizes the urgent times we live in as an era of unprecedented planetary ecocrisis that threatens the extinction of all forms of life on Earth (Best and Nocella II, 2006). Kahn (2010) powerfully challenges educators throughout the humanities and social disciplines to make eco-literacy the center of their discussions. He asserts this as a moral challenge that helps students to reflect upon Western industrial human misuse of natural resources, exploitation, and violence toward other species. Given the magnitude of such ecological catastrophe, conversations are increasing and many people are raising this problem in the hope of transforming societies. Hence, governments, non-governmental organizations, and some private institutions have frequently emphasized environmental education. Kahn (2010) points out that a major problem, however, is that ecological issues are generally taught only within the natural sciences departments across the universities. And these degrees often do not train students in eco-literacy, preventing critical thought and analyses of the politics and ideologies of our societies regarding environmental issues (pp. 6–7). Kahn (2010) offers engaging foundational suggestions to those aware of the importance of the ecopedagogical movement:

> Provide openings for the radicalization and proliferation of ecoliteracy programs both within schools and society; Create liberatory opportunities for building alliances of praxis between scholars and the public (especially activists) on ecopedagogical interests; Foment critical dialogue and self-reflective solidarity across the multitude of groups that make up the educational left during an extraordinary time of extremely dangerous planetary crisis. (pp. 27–28)

Kahn's call for critical ecopedagogical approaches in the social sciences is long overdue; otherwise, our current planetary crisis would not be as severe as it is. Global Education in TFL can and should incorporate a critical ecopedagogy approach as it advocates for an education that addresses global justice, human rights, sustainability, peace, and intercultural communication, and it must be based on tolerance, solidarity, equality, justice, inclusion, cooperation, and nonviolence (Global Education Network, 2009). Both Global Education and critical ecopedagogy call for a sustainable, harmonious, and nonviolent world. Such a world must take into consideration and respect the lives of all species on it.

Human history reveals that many cultures have exploited the lives of nonhuman animals, but currently such exploitation has reached an unprecedented level in regard to what happens to nonhuman animals. Factory farm confinement, where animals are crammed into high-volume "production facilities," living in cages that might be better described as torture devices (as they cannot even move their limbs) is a vile common practice both of the first world and of many emerging and developing countries. More examples of nonhuman animal exploitation are zoos that pass along the conservationist talk but in reality imprison animals and expose them to the public for profit, and the fur industry where animals are trapped in bone-breaking steel traps, clubbed to death, electrocuted through the mouth or anus, have their necks broken, and sometimes are skinned alive. An important part of ecopedagogy education is addressing and challenging the types of violence nonhuman animals suffer. Tom Reagan's (2004) essay "How to Justify Violence" gives a glimpse of what animals used by humans have to endure:

> Animals are drowned, suffocated, and starved to death; they have their limbs severed and their organs crushed; they are burned, exposed to radiation, and used in experimental surgeries, they are shocked, raised in isolation, exposed to weapons of mass destructions, and rendered blind or paralyzed; they are given heart attacks, ulcers, paralysis, and seizures; they are forced to inhale tobacco smoke, drink alcohol, and ingest various drugs, such as heroin and cocaine. (Reagan, 2004, p. 235)

While several studies have been carried out that focus on environmental education within Global Education in TFL, investigations that report on nonhuman animal exploitation are, to date, nonexistent. Currently, prominent foreign language curriculum theory and instructional practices do not offer a real challenge in questioning cultural practices of nonhuman animal exploitation, normalization of animal abuse, and the institutionalization and language of speciesism. Therefore, these approaches are in need of a "radical reconstruction away from an uncritical form of environment literacy" that has not challenged standard forms and cultural practices of animal oppression (Kahn, 2010, p. 27).

As an educator in the field of foreign languages, and as a recent investigator and activist on nonhuman animal issues, I attempt here to start a conversation on how foreign languages curriculum theory and instructional practices must be revised in order to envision TFL fundamentally grounded upon eco-literacy—major curricular and pedagogical shift in TFL to bring animal issues from marginality to the center of today's most critical social injustices. In order to take steps toward such a reconceptualization, TFL curriculum must be adjusted in order to create opportunities between scholars and students, but also with the public (especially activists) to make coalitions of praxis regarding ecopedagogical interests, including nonhuman animal advocacy and liberation. Radical pedagogy of animal liberation also compels instructors to be activists, besides mentors; or, at least, it expects that academics collaborate with activists, as these two viewpoints combined are an effective and urgent path to promoting awareness for the plight of nonhuman animals and, consequently, to attain their liberation. Activists play a crucial role in animal liberation as they physically fight to liberate animals from legalized institutions that keep them in prison. Bernstein (2004) explains that activists fight a legitimate war, but without weapons, without harm, and with the purpose of only liberating nonhuman animals from domination and pain (Bernstein in Best & Nocella, 2004, p. 101). However, it is all too often that the government and the media label activists as 'terrorists', which prevents dialogue and collaboration between academics and activists. Nocella (2004) calls for an alliance between academics and activists; he argues that a cohesive support from academics would help activists increase legitimacy for their actions; however, before this alliance is built, it is necessary to develop a "method that academics and others can use to understand a complex political movement" (Best & Nocella, p. 196). Although this article is not a direct call for instructors of foreign languages to engage in physical acts of liberating animals, or even to become part of the Animal Liberation Front movement, I am asserting that an ecopedagogical movement requires the inclusion of a clear and critical discussion of animal oppression. Thus, a discipline committed to eco-literacy requires and empowers instructors of foreign languages to teach with courage, to speak out about the horrors and atrocities committed to animals, to engage in effective animal advocacy, and to make alliances and cooperate with local and national animal rights activists.

## *Developing Ideas on Bringing the Animal to TFL*

Critical pedagogy and Global Education practices in TFL curriculum must be revised and modified into ecopedagogical liberationist approaches that make the teaching of nonhuman animal issues a priority. The beginning is not a laborious task! Some of the most common practices used in TFL are already suited to bringing awareness of nonhuman animal exploitation, such as role-playing, projects, and service learning, to name a few.

Role-plays encourage students to imagine themselves as different people, in different situations, providing connection between the classroom and world beyond (Livingstone, 1983 cf. Swartz). The effectiveness of role-plays as an educative tool to engender critical thinking and reflective praxis is robust. Role-play scenarios are abundant. For example, in the lower levels of language instruction, students are often engaged in a type of role-play in which a pair works a dialogue, a sketch, an interview, and so on; the pair works on their language proficiency while opining about a chosen subject that can vary from social problems, cultural differences, entertainment, sports, and so on. Interviews of famous people, athletes, politicians, roommates, and family members are a common role-play practice often engaging students.

An example of a foreign language ecopedagogical curriculum engaged with nonhuman animal advocacy could help students create an interview in which the interviewee is an animal rights activist. Prior to the activity, students would need to be familiarized with some basic nonhuman animals issues and the types of works activists do. To recreate an animal rights activist milieu, for example, students could engage in peer meditation skills, strong critical thinking, and reconsideration of what many, to use the words of Best and Nocella (2004), "perceive as their absolute God-given rights to exploit animals by sole virtue of their human status" (Best & Nocella, p. 13).

Project-based learning increases critical thinking skills and fosters positive attitudes toward varied subjects (Beckett & Miller, 2006). Projects can be individual or collaborative enterprises and are carefully planned to achieve a particular aim. An example of a final semester interdisciplinary project would be to engage students in working with nonhuman animal advocacy groups based in the countries of their target language. Subareas, either chosen by students or advised by instructors, would include working with animal shelters, sanctuaries, endangered species conservation groups, food justice, and so forth. Such a project, without immersion, could still be successfully accomplished through an exhaustive and guided use of technological tools, such as the use of Tandem Language Learning (Retrieved March 12, 2017, from https://en.wikipedia.org/wiki/Tandem_language_learning) in which students can be paired up to interact specifically with native speakers concerned and working with the same issues.

Service learning projects help motivate students in their foreign language learning by giving them an opportunity to use their skills in serving others. Service learning is action and reflection integrated with academic curriculum to enhance student learning and to meet community needs. Investigations and practices for Service Learning in the teaching of Spanish as a foreign language in the United States, for example, show optimistic results regarding students' motivation and also the positive effects of both language learning and social issues understanding (Caldwell, 2007; Weldon & Trautmann, 2003). The praxis of critical service learning pedagogy, drawn from Paulo Freire's ideology, must emphasize cultural action for freedom, a balance of power and conscientization. An important aspect of critical service learning is dialogue. The nonhierarchical nature of dialogue is fundamental in Freire's pedagogy. In a Freirean approach, dialogue occurs between subjects who are open to seeing the world through the eyes of others. Humans cannot engage in a dialogue with animals; the horizontal nature of language in this case is invalidated. But the task must be to create a dialogue about nonhuman animals based on a "profound love for people and for *the world*" (Freire, 1972, p. 70).

While there are many opportunities for critical service learning within their own communities, students of foreign languages benefit from it thoroughly when they experience cultural immersion. Students who participate in SAPs usually find that they to have become more proficient in the target language, and more open to intercultural communication; students believe that the experience of living abroad and communicating in the target language helps them understand their "own cultural values and biases," and it also helps them to "develop a more sophisticated way of looking the world" (Dwyer & Peters, 2004). SAPs are robust opportunities to create nonhuman animal advocacy as students can work in direct contact with local activists, rescuers, nongovernmental and governmental organizations, either by volunteering or carrying out ethnographic interviews as long as ecopedagogical SAPs are carefully modeled to become a praxis of animal liberation.

## *The Task of Revising Foreign Languages Curriculum and Educating the System: A Radical Pedagogy Approach*

Foreign Languages course material, including textbooks, often carries, uncritically, the language of the oppressor, speciesism, that is, mainstream didactic teachings of cultural practices based on violence and exploitation of animals. For example, Spanish and Portuguese textbooks frequently depict a controversial practice, bullfighting, as a cultural tradition. Too often these depictions disregard several anti-bullfighting groups and the work of many activists—every year thousands of Spaniards take to the streets of Madrid to protest against bullfighting—and fail to engage students on a discussion about bullfighting lobbying, the major force that drives this practice, according to Gálvez (2016). Other forms of animal exploitation, common in major cities of foreign countries, such as zoos, aquariums, and rodeo arenas, also often pass without criticism in mainstream course material; instead, they are presented as a cultural norm.

Textbook chapters on food vocabulary depict animals as what Carol Adams (1990) called the "absent referent" (p. 65), that is, the flesh that is served up at the table is not perceived to be related to the animal that moves about in the field; animals are consumed as food without any reference to their past, their living beings.

> Animal's lives precede and enable the existence of meat…without animals there would be not meat eating, yet they are absent from the act of eating meat because they have been transformed into food. (Adams, 1990, p. 65)

The next point Adams makes offers exciting points for potential investigations into language theories, and first and second language acquisition. Future investigations in this subject can potentially contribute to ethical discussions about how languages perpetuate animal exploitation and how cross-linguistic awareness can help foster nonhuman animal empathy. Adams (1990) argues that, "animals are made absent through language that renames dead bodies… language contributes even further to animals' absences" (p. 65).

For now, there is room to initiate a discussion on visual literacy in TFL textbooks and other materials regarding food vocabulary, which may ethically challenge this naturalized practice of absolute human dominion over nonhuman animals. According to the Humane Society of the United States about 4.6 billion animals (cattle, hogs, chickens, sheep and lambs, and turkeys) were killed in 2015, and this data does not include the slaughter of fish. Considering this abysmal slaughtering of animals for food, the subject of food vocabulary in foreign languages teaching must focus on ecopedagogical approaches to address this horrifying social injustice; there must be opportunities for dialectical conversations between educators, language supervisors, and textbook authors, which, optimistically, will crown toward "building alliances of praxis between scholars…on ecopedagogical interests," providing "openings for eco-literacy programs" and stimulating "critical dialogue and self-reflective solidarity across multitude of groups…during an extraordinary time of extremely dangerous planetary crisis" (Kahn, 2010, pp. 27–28). There is this story instructors of Portuguese as a foreign language (PFL) often share during lunch-break. This story is, in reality, an optimal opportunity for TFL educators to work within an eco-literacy lens. In the first semester of PFL, students early on learn adjectives, one of which is *quente* (hot). In about the middle of the semester, students learn vocabulary for the name of animals, one of which is *cachorro* (puppy). Toward the end of the semester, students, then, learn vocabulary for food, one of which is *cachorro-quente* (hot-dog, or *hot-puppy* literally). While in first language acquisition the association is not well developed, (first language acquisition is mostly based on repetition) in second language acquisition (especially when learners are adults) the association is fully developed. So, when students of PFL are introduced to the word *cachorro-quente* (hot-dog) they demonstrate shocking reactions: amusement, sadness, and discomfort. It is as only in the target language that they come to think about the literal meaning of the word "hot-dog." And although it is not a dog that is cooked, in the students' cultural background in the students' cultural background puppies are not seen as food quite the opposite, for many, they are part of the family. This situation becomes a fertile ground to engage students in a dialogue about why we love some animals but eat others arbitrarily as asked Melanie Joy (2009).

I have been constantly vigilant since I became involved with nonhuman animal issues as the normalization and desensitization of animal exploitation are presented in every single subdivision of Western industrial society, including and especially within our own educational institutions. Radical change often comes with the works of activists who sometimes put their lives at risk to liberate animals from their current imprisoned state. But one does not necessarily need to risk their lives to advocate for nonhuman animals; there is much we, as educators and global citizens, can do. The opportunities for engaging on animal advocacy with other academics, across disciplines and across departments, are robust. A couple of years ago, I opened up the website of an American college and navigated to its Department of Spanish and Portuguese. The background of the website depicted a painting of a pompous bullfighter throwing his red cape into a bull's face covered in blood. I e-mailed the Spanish professor, responsible for arranging the design of the department's webpage, saying that I found the painting contentious. He argued that the image of the bullfight in no way constituted an endorsement of the practice but rather it was simply a piece of art worthy of reflection, and that he, himself, opposed the practice. I argued back asking if the works of many animal rights activists against bullfighting (some of which throw themselves into arenas in protest, putting their lives in danger) should be completely ignored if they kept the painting. He took the matter to the executive committee of the department and the background painting was, then, replaced by one of a flamenco performer. In another American language department, a flyer advertising Study Abroad courses for French students in Morocco showed a picture of one of their former students holding a chained macaque in the streets of Marrakesh. The human who "owns" the macaque apparently let the student take a selfie, which, in turn, was then used to promote the department's French Study Abroad Program Morocco. Once again, I engaged in a conversation with the head of the department about the problematic representations of such a picture. In contrast to the Spanish professor and the website example, this person in a position of power immediately agreed with my observation and mentioned how he did not think about it, concluding "but now that you're saying, it makes totally sense." The flyer was replaced. The reason I mention the above examples is to emphasize the normalization and desensitization of animal abuse, and how complex it is to engage in educating and liberating the many segments of our societies, including our own educational institutions. But it is also a powerful reminder that dialogue transforms societies (Freire, 1972).

## Conclusion

A growing number of language-teaching professionals are finding that Global Education presents an exciting approach to their work, which can promote global awareness, international understanding of foreign cultures and peoples, and a commitment to working toward a better world. Global Education in TFL is becoming an important part of language theory and curriculum practices. Students who learn a foreign language these days become not only proficient and able to communicate with other cultures, they also become more aware of global problems and social injustices. Global Education and interdisciplinary foreign language curricula help students engage with issues regarding human rights, peace education, intercultural communication, socioeconomic and poverty, among others.

Environmental issues in Global Education usually focus on problems related to pollution, deforestation, recycling, and natural disasters; while also discussing global warming and associated social injustice but rarely, if ever, do such lessons directly address animal exploitation and its damage to the environment and connections to unjust social suffering—especially by explicitly engaging with critically understanding agriculture as a key part of the animal industrial complex. Ultimately, how educators teach social and environmental issues in Global Education is in need of radical revision. The exponential number of the human population and humanity's unprecedented use of animals to satisfy their needs is unsustainable, and it is also the source of the greatest social injustice of our times, paraphrasing the words of Phillip Wollen (2012). Global Education in TFL must make the topic of animal exploitation a priority by working with eco-literacy and practices into language curricula across the country. In this article I offered my own explanatory attempt to justify the importance of an inclusive ecopedagogical approach in TFL that places the social injustice of nonhuman animal exploitation in the center as the magnitude of the current ecological crisis, to which animal exploitation—especially animal agriculture—is responsible, and also the unnecessary suffering and violence perpetrated to animals which can no longer be marginalized in our classrooms. This task is complex; the literature (both theoretical and practical) is insufficient; and the resistance is immense, as animal use is grounded in deeply rooted cultural traditions. As educators and scholars of foreign languages committed to critical thinking and Global/ecopedagogies, we must initiate dialogues to revise course materials, develop ecopedagogical approaches, build interdisciplinary alliances with other educators/academics/local and national activists, and engage on research. Our field of instruction, foreign languages, is an undoubtedly fertile grounds to teach nonhuman animal issues as inextricable from social justice and sustainability for global communities.

## *References*

Adams, C. J. (1990). *The sexual politics of meat: A feminist-vegetarian critical theory* (20th Anniversary Edition). New York, NY: Continuum.

Beckett, G. H., & Miller, P. C. (2006). *Project-based second and foreign language education: Past, present, and future.* Greenwich, CT: Information Age Publishing.

Berkhout, F., Leach, M., & Scoones, I. (2003). Shifting perspectives in environmental social science. In *Negotiating environmental change: New perspectives from social science* (pp. 1–31). Cheltenham: Edward Elgar.

Bernstein, M. (2004). Legitimatizing liberation. In S. Best & A. J. Nocella II (Eds.), *Terrorists or freedom fighters? Reflections on the liberation of animals.* New York, NY: Lantern Books.

Best, S., & Nocella, A., II. (Eds.). (2004). *Terrorists or freedom fighters? Reflections on the liberation of animals.* New York, NY: Lantern Books.

Caldwell, W. (2007), Taking Spanish outside the box: A model for integrating service learning into foreign language study. *Foreign Language Annals*, *40*, 463–471.

Correa, M. (2011). Advocating for critical pedagogical approaches to teaching Spanish as a heritage language: Some considerations. *Foreign Language Annals*, *44*(2), 30.

Crawford, L. M. (1978). Paulo Freire's philosophy: Derivation of curricular principles and their application to second language curriculum design (Unpublished doctoral dissertation). University of Minnesota.

Crookes, G. (2012, November 5). Critical pedagogy in language teaching. Retrieved from http://onlinelibrary.wiley.com/doi/10.1002/9781405198431.wbeal0284/pdf

Darder, A., Baltodano, M., & Torres, R. D. (2003). *The critical pedagogy reader.* New York, NY: RoutledgeFalmer.

Della Chiesa, B., Scott, J., & Hinton C. (Eds.). (2012). *Languages in a global world: Learning for better cultural understanding.* Paris, France: OECD Publishing. doi:10.1787/9789264123557-en

Dwyer, M. M., & Peters, C. K. (2004). The benefits of study abroad. *Transitions Abroad*, *37*(5), 56–58.

Gálvez, J. J. (2016, April 8). Pro-bullfighting lobby throws economic data into the ring. *El País.* Retrieved from http://elpais.com/elpais/2016/04/06/inenglish/1459937034_734626.html
Freire, P. (1972). *Pedagogy of the oppressed.* New York, NY: Herder and Herder.
Fujikane, H. (2003). Approaches to global education in the United States, the United Kingdom and Japan. *International Review of Education/Internationale Zeitschrift Für Erziehungswissenschaft / Revue Internationale De L'Education, 49*(1/2), 133–152. Retrieved from http://www.jstor.org/stable/3445478
Guha, R. (2000). *Environmentalism: A global history.* New Delhi: Oxford University Press.
Jacobs, H. H. (Ed.). (2010). *Curriculum 21: Essential education for a changing world.* Alexandria, VA: Association for Supervision and Curriculum Development.
Kahn, R. V. (2010). *Critical pedagogy, ecoliteracy and planetary crisis. The ecopedagogical movement.* New York, NY. Peter Lang Publishing.
Kindness Trust. (2012, May 16). *Phillip Wollen: Animals should be off the menu debate.* Retrieved from https://youtu.be/uQCe4qEexjc
Maastricht Declaration on Global Education in Europe. (2003). A European Strategy Framework
For Improving and Increasing Global Education in Europe to the Year 2015. Global Education Nework Europe. Retrieved from https://gene.eu/wp-content/uploads/Gene_EuropeanPolicyFrameworks-TheMaastrichtDeclaration2002.pdf
Magnin, M. C. (2002). An interdisciplinary approach to teaching foreign languages with global and functional simulations. *Simulation & Gaming, 33*(3), 395–399.
McLaren, P. (1989). *Life in schools: An introduction to Critical Pedagogy in the foundations of education.* London, UK: Longman
Millennium Ecosystem Assessment. (MEA) (2005). Retrieved from http://www.millenniumassessment.org/en/index.html
Nocella, A. J., II. (2004). Understanding the ALF: From critical analysis to critical pedagogy. In S. Best & A. Nocella (Eds.), *Terrorists or freedom fighters? Reflections on the liberation of Animals* (1st ed.). New York, NY: Lantern Books.
Plumwood, V. (1996). Androcentrism and anthrocentrism: Parallels and politics. *Ethics and the Environment, 1*(2), 119–152. Retrieved from http://www.jstor.org/stable/27766018
Reagan, T. (2004). How to justify violence. In S. Best & A. Nocella (Eds.), *Terrorists or freedom fighters? Reflections on the liberation of animals* (1st ed.). New York, NY: Lantern Books.
Riasiti, M., & Mollaei, F. (2012). Critical pedagogy and language learning. *International Journal of Humanities and Social Science, 21*, 223–229.
Ross, E. W. (2016). Life in schools: An introduction to critical pedagogy in the foundations of education by Peter McLaren. In J. L. DeVitis (Ed.), *Popular educational classics: A reader.* New York, NY: Peter Lang.
Swartz, J. J. (n.d.). Carol Livingstone: Role play in language learning. Retrieved from http://dx.doi.org/10.5785/1-2-507
Weldon, A., & Trautmann, G. (2003). Spanish and service-learning: Pedagogy and praxis. *Hispania, 86*(3), 574–585.
Yakovchuk, N. (2004). Global issues and global values in foreign language education: Selection and awareness-raising. *English Language Teacher Education and Development, 8*, 28–47.

# *Part V: Activism*

# 9. *Activist Education for Animal Ethics: The Imperative of Intervention in Education on the Non/Human*

J.L. Schatz

Teaching total liberation requires more than mere theoretical discussions and isolated instances of activism. It requires grading on more than just what is said in the classroom or turned in at the end of a semester. It requires incentivizing the necessary lifestyle changes that translate theoretical foundations into concrete practices. Feminists have long argued that educators must refuse to divide the public from private in one's approach to politics and education (Peterson, 2000). Following in this tradition, my chapter argues that evaluating students beyond what is traditionally considered gradable in the public sphere is not only the most effective way to teach non/human liberation but also one of the quickest ways to achieve its realization. I contend that such teaching practices are not only more effective in producing change but that they also foster a more enriching experience for students and teachers alike. A foundational premise of my chapter is that speciesism is every bit as insidious as any long list of –isms that exist (Best, 2013). Challenging speciesism within the academy is essential to defeat it since the rigorous testing of contemporary norms through classroom debate can dislodge the hegemonic assumptions about what is to count as fully human. To do so, student and teachers must take speciesism "seriously, not simply as another interesting category for academic analysis but as an ethical and political imperative. [This] "...implies doing everything we can to dismantle the linguistic, conceptual and performative apparatus which makes all kinds of violence and oppression possible" (Dell'Aversano, 2009, p. 100). As educators, there is imperative to not reward students who benefit from actions that are outwardly sexist, racist, heterosexist, or otherwise violent in or out of classroom spaces. The same must be true for speciesism if we are to bring about the existence of non/human liberation. In all instances, educators must be willing to grade in a way consistent with the subject material they teach.

When I began teaching units on speciesism in 2001 at the undergraduate level at Binghamton University, the majority of students I encountered were unaware of the horrors of factory farms. At that time, I was teaching Peter Singer's (2001) *Animal Liberation*, which gives an excellent overview of factory farming's deplorable conditions and the unnecessary torture animals are forced to endure in the name of profit, and Carol Adams' (1990) *The Sexual Politics of Meat*, which explains the interconnection between a heterosexist form of patriarchy with speciesism and the consumption of flesh. Even less were aware about the deplorable working conditions and environmental devastation that goes alongside the slaughter at industrialized killing facilities (The Guardian, 2010). Over my fifteen years of teaching, I have since witnessed increasing number of students enter my classroom already aware of the horrors of factory farming, leather production, and animal testing. When I ask now in 2016, there is at most a small handful of students who haven't seen documentaries like PETA's *Meet Your Meat* or *Food Inc.* (2008). Almost no one in my classes are willing to defend what happens in factory farms as necessary or ethical. Those who do are students who are wholly unwilling to take the lives of non/humans into moral consideration. These students tend to believe that no action against non/human animals could be unethical. However, again, I encounter very few students who fully believe that non/human animals have no intrinsic value whatsoever beyond the status of objects. Despite this change in awareness, there has not been a significantly higher percentage of students who are vegetarian or vegan enrolled in my classes. There isn't even a larger number of localtarians or people who go out of their way to buy family farmed animals. Sadly, the majority of students remain content buying flesh off the fast food dollar menu on their way home.

The question I ask during my speciesism units and outside of class is then: Why? Why continue to pay for something you know is wrong? Why willingly act in ways that are unethical by your own admission? The most popular two answers I receive is the belief that not eating animals won't do anything to actually stop the industry, and that it would be too hard to actually go vegetarian in practice. Semester after semester the handful of vegetarians or vegans would convince people that, in the university setting, it's increasingly easy to abstain from eating animals. They would also point out how one's ethical responsibility is about oneself; not whether one can change the entirety of society. In short, refusing to pay for the torture or slaughter of non/human animals should be an imperative even if it doesn't bring an end to speciesism.

Put simply, it is about the individual's moment of the encounter that determines ethics and not a universalized endpoint of change. To this end, when comparing ethics to ontology, Pat Gehrke (2010) contends, "Responsibility is first and foremost found in my ability to respond, which is given to me by the Other. If I respond with non-response then I do even more profound violence to my responsibility and to the Other." Unfortunately, when I asked about these moments of awareness if students would start consciously limiting their factory farmed consumption, many were still unwilling to commit. In their heads, if they couldn't give it up entirely why bother to try at all. At the same time, they seemed to recognize how that argument was more of an excuse than a reason not to try. Even still, people did not change their actions.

As a result, in 2012 I decided to implement an experiment I called the veg pledge. At the conclusion of teaching my unit on non/human liberation, I would offer students anywhere between one to six extra-credit points on their final paper for committing to going vegetarian or vegan for the remainder of the semester; or through winter break depending on where the unit fell. Students who were already vegan had to figure out something equivalent in scope that they consumed and didn't abstain from previously—whether it be 100% cruelty-free shopping, boycotting sweatshop labor, abstaining from other forms of institutionalized injustice, or something else altogether was up to the individual student. Students who might have dietary or medical conditions did not have to disclose such preventative issues. Instead they could commit to giving up some other form of speciesist violence, including but not limited to kinds of soap, household cleaners, makeup, or any number of other things they habitually purchased. I would also send out a class e-mail with links instructing students on how to make the shift in a healthy way, and how to find the necessary resources to research consumer brands. Beyond that I would encourage students to conduct additional research into their specific dietary needs and made myself available whenever needed. At the end of the semester students who chose to participate were asked to submit a short statement of 300–500 words describing their experience. They were asked to explain how many extra-credit points they felt they deserved based upon their honest effort to abstain from institutionalized speciesism in their everyday lives.

Overwhelmingly most students not only enjoyed the experience but were honest about those instances where they slipped up and ate animal products in particular instances. Over half of the fifty-six statements I received, during the course of four semesters, admitted to having ate animals out of momentarily forgetting their pledge, not knowing a certain product contained animal flesh, or social pressure by family or friends at a particular event. Only two gave up entirely after repeatedly messing up. The rest described the ongoing struggle to try and limit their consumption. I take these testaments, along with the fact that only about two-thirds of my students decided to take the pledge, as evidence of student honesty in reporting. Twenty-four students noted how it was much easier than they thought it would be and another thirteen noted that after getting past an initial learning curve it was easier than they thought. Only twelve said the process was hard, and several each semester said they planned on continuing to be vegetarian after the class ended. A large percentage of those who gave up on the dietary portion of the pledge shifted to other forms of cruelty-free shopping to at least do something.

I quickly realized that this external reward for implementing course material in everyday life was more effective at convincing students to live up to their own beliefs against factory farms, and other forms of exploitation, more than anything that could be taught in the classroom alone. Students found themselves empowered about their dietary choices as they discovered more about the choices they made in relation to their everyday consumption practices. They learned hands-on the extent that speciesism permeates Western capitalist institutions. One student wrote on her experience,

> With each trip to the store, I made the effort to verify that what I am buying is animal-friendly, as well as preferably free-trade. My decisions have made me feel really content as a person as well as a student; my attempt at extra-credit turned into a form of paying it forward. Your generosity initiated my generosity to the companies that treat all lives as precious, regardless of animal or human. I plan to continue this effort, particularly because I enjoy diverging my habitual way of shopping to try new and more ethical products.

Several students continue to remain in touch with me and have become increasingly activist in their challenging of speciesism as a result of taking this first step. Many have helped their friends and family realize the possibility of giving up eating animals through simply modifying their practices as a result of the pledge. Ultimately, the experiment was wildly more successful than I could have ever imagined even though the extra-credit amounted to at most a third of a letter-grade on their final paper, which was only a small percentage of their overall grade.

Despite the success I've had, when I talk with my colleagues several decry the coercive nature of my experiment that culminates in a biopolitical self-monitoring of what students consume. Many who criticize my experiment argue that it is a subjective expression of the tyrannical whims of my desires, which some see as an abuse of power. This became evident the first time I tried to implement a veg pledge paradigm in the intercollegiate debate community before bringing it to the classroom (CEDA Forums, 2012). They would note that if I could implement a veg pledge then there would be nothing to stop another instructor to give extra-credit to students who pledge to hunt, murder, or anything else. In the first instance they would claim that such teaching methods are a subjective imposition of my ethics onto a student's life. This argument falsely valorizes a mode of passive liberal teaching where instructors are assumed to be entirely objective in a student's pursuit of knowledge, while respecting all arguments and positions as equal (Schatz, 2013). Further, such beliefs ignore not only how such objectivity is impossible but also how they are undesirable since all arguments are not equal in merit. Racism, sexism, and speciesism are among a long list of ideologies that should not be tolerated within classroom spaces out of the pretense of remaining objective. No doubt,

> objectivity...has long been seen by feminists as gendered, structuring what counts as knowledge[,] ...which is how it has been defined by particular kinds of people who are male, white, and middle class. An example of such feminist work is that of Donna Haraway, who argues that objectivity was associated in the eighteenth century with English gentlemanly masculinity and his "enlightenment" quest for knowledge... [that] was seen as "transparent," independent, rational, and honorable, quite the opposite of the "feminine" and those who were seen as "dark," "uncivil," and "unruly." (Swan, 2010)

An objective instructor or neutral adjudicator necessarily privileges dominant knowledges and subject positions because they refuse to intervene to create an environment for subordinated ideas and practices to have an equal opportunity to flourish. Absent intervention dominant performances, which are already hegemonic in scope, will continue to limit the means of expression and the possibility for resistance. For students to understand an opposing viewpoint they must be made to truly see the opposition's argument, which requires a living out of performances, and not just academic role-playing.

Sadly, this notion of activist intervention runs counter to the liberal model of free speech that has promoted a form of moral relativism in academic spaces. Students who can write a well-written paper justifying the basis of slavery should now not only be tolerated but rewarded if they can demonstrate their skill to argue well, saying nothing of the ethical merits of the arguments they're writing.

> In other words...what is proclaimed and practiced as tolerance today, is in many of its most effective manifestations serving the cause of oppression. ... The elimination of violence, and the reduction of suppression to the extent required for protecting man [*sic*] and animals from cruelty and aggression are preconditions for the creation of a humane society. ... As deterrents against nuclear war, as police action against subversion, as technical aid in the fight against imperialism

> and communism, as methods of pacification in neo-colonial massacres, violence and suppression are promulgated, practiced, and defended by democratic and authoritarian governments alike, and the people subjected to these governments are educated to sustain such practices as necessary for the preservation of the status quo. Tolerance is extended to policies, conditions, and modes of behavior which should not be tolerated because they are impeding, if not destroying, the chances of creating an existence without fear and misery. (Marcuse, 1965, p. 1)

It is the burden of an educator to remap the notion of tolerance so that students are encouraged to withdraw from their passive acquiescence to institutions of violence they disagree with in theoretical classroom discussions of oppression. When students are academically opposed to racism, sexism, homophobia, speciesism, or any other –isms it should be the obligation of the instructor to inspire that student to be empowered to live out what they believe and make a difference. Finding ways to encourage students who disagree with factory farming practices to give up paying for its continuation provides the necessary tolerance for a radical vegetarian politics to be realized.

What is crucial to remember here is that students are encouraged to live out those things they already agree with versus forcing them to go against their morals. This fact makes the veg pledge model of extra-credit very different from incentivizing students to murder, rape, break laws, or do something against their moral beliefs. Rather, it is a model of education that jolts students out of their complacency to immediately do something with the education they obtain in the classroom. "As the critical educator Paulo Freire remarked, education is not itself the lever of social change but it can play an important role to the degree that it works curricularly [*sic*] to generate counterhegemonic knowledge and stir the feelings of socio-political protest in students" (Kahn, 2009). Until more instructors empower students to internalize what they're learning, it will be impossible to break through the passivity that has been bred into current educational models that presume the instructor should be entirely neutral. Some things are wrong, and instructors should not be afraid to call them out as such. Factory farms are one of those very things since there is no moral or ethical justification for their continued existence. To say this is not to say that every time an individual consumes factory farmed flesh that such an individual is unethical. To be sure, socioeconomic conditions built off of years of racially implemented capitalist structures have made the unhealthy 99 cent chicken nugget the only option for a large number of people (Harper, 2010). Nevertheless, at most state universities in the United States, the vast majority of students have the ability to be vegetarian in a healthy way at a similar cost as what they pay for food already. It is just a question of the knowledge to know how.

Teaching students how by rewarding them for exploring ethical and political possibilities outside the classroom promotes an empowering new form of objectivity. Students can now take those moments where they objectively know they can and should pass over animal products and act upon them instead of brushing aside those instances as merely an academic argument. Donna Haraway writes,

> "feminist objectivity means quite simply situated knowledges" ... [that] helps students learn to "situate" the assumptions, limitations, and political and social influences of any particular research or creative project. In order to do so, students need to explore gaps in the research questions and representational techniques the traditional disciplines have adopted; to learn alternative cultural reading and viewing strategies; and to incorporate into the center of feminist analysis questions concerning epistemology. (Pryse, 2000, p. 115)

It is only through the willingness to explore the gaps where traditional modes of grading ends that students and instructors can discover new educational mechanisms of learning that goes beyond the classroom. To teach students how to realize their ethics and beliefs beyond a mere paper is one of the more profound things that can take place during a semester since by nature the material becomes about more than just the class alone. As a consequence of this model, "the reversal of the trend in the educational enterprise...could conceivably be enforced by the students and teachers themselves... [by] the systematic withdrawal of tolerance toward regressive...opinions and movements... [through] resistance at particular occasions...at the local and small-group level" (Marcuse, 1965, p. 7). Ultimately, this produces a new objective awareness over what one performs outside the classroom that impacts and enriches one's subjective comprehension of the material that leads to tangible change.

Such awareness radically rethinks what is traditionally taken as academia in a positive direction for three distinct reasons. First, it takes the university as something more than a training ground for future jobs, graduate specialties, or avenues of activism by making the process of learning part of the end product as well. To put it simply, the university classroom is an end in of itself as much as it is a means to some future tomorrow. Failure to understand it that way disempowers the educational opportunity that exists in those moments of learning. This is why "activists must therefore stick relentlessly to exposing cruelty on farms, and seeking out the audiences who are most likely to respond [in] ...high schools and colleges" (Marcus, 2005, p. 123). Second, it puts an active responsibility on teachers and students alike to become critical of their consumption habits both in terms of education and in the flesh. Absent this responsibility education turns into empty signifiers of excellence while producing a vacuous form of learning that is disempowered through becoming a matter of rote repetition (Spanos, 1993). As bell hooks (2006) points out,

> Stuart and Elizabeth Ewen emphasize in *Channels of Desire*: The politics of consumption must be understood as something more than what to buy, or even what to boycott. Consumption is a social relationship, the dominant relationship in our society–one that makes it harder and harder for people to hold together, to create community. At a time when for many of us the possibility of meaningful change seems to elude our grasp, it is a question of immense social and political proportions. To establish popular initiative, consumerism must be transcended–a difficult but central task facing all people who still seek a better way of life. (p. 376)

Having students merely reading the words of theorists in abstract, instead of meaningfully applying their words outside of class or term papers uncritically consumes the power of academic theorization. To teach (students) critics like Carol Adams (1990), Marjorie Spiegel (1997), Steven Best (2013), or any other number of animal liberationists requires taking the subject of oppression as something more than mere academic analysis. To do so merely reobjectifies the slaughtered bodies of non/human animals to be dissected and deconstructed in dissertation after dissertation. Academic consumption, just like every other form, is always already a sociopolitical act that can never be neutral.

The third reason veg pledge models of education positively shapes academic encounters is that it deconstructs the belief that teaching must be fully impartial and objective by exposing this assumption as both unrealistic and dangerously undesirable. To teach purely from a neural and objective vantage point removes teaching from the equation by tolerating any and every act of discursive performance of oppression as equal. This argument has been pointed out by critical race theorists who have long warned about teaching and learning as a detached observer. To this end, George Yancey (2005) criticizes intellectuals who take the

> embodied self… [as] irrelevant to theory, superfluous and cumbersome in one's search for truth. It is best, or so we are told, to reason from nowhere. Hence, the white philosopher/author presumes to speak for all of 'us' without the slightest mention of his or her 'raced' identity. … It is important to note that this particular strategy also functions as a lens through which to theorize and critique whiteness; for the Black body's 'racial' experience is fundamentally linked to the oppressive modalities of the 'raced' white body. (p. 216)

As such, the appeal to neutrality misunderstands current university practices as already somehow unbiased in its institutionalized quest for excellence that grades students in normative ways. To require students to always write about certain topics, or conform to only the confirmed notions that instructors present as truth in multiple choice tests, is already every bit as subjectively confining to students' thoughts and practices. Instead instructors should actively intervene in order to educate a change in both theory and praxis while encouraging students to live out their educational endeavors in their everyday lives.

Thus, unlike the veg pledge models I've been experimenting with, these so-called objective models don't encourage students to reflect on their participation in the learning process. As a result, students become disempowered and the education potential lost. They are taught only how to repeat the arguments and performances they believe the teacher wants to hear instead of internalizing the material. Neutral teaching practices also cause instructors to suppress a part of themselves to retain a guise of objectivity instead of being able to teach from their fundamental beliefs on the topics they're experts on. Teaching animal liberation, feminist, or antiracist theory requires a position of non-neutrality. "Racial and sexual discrimination are *always* wrong. ... It is not a reformed, 'more humane' rape that an enlightened ethic calls for; it is the abolition of all rape that is required; it is this extreme position we must uphold" (Regan, 2013). To make a student think like a feminist but do nothing to educate them on how to fight patriarchy will never be enough. Likewise, to get students to agree that speciesism is wrong and that factory farming is both immoral and ecologically suicidal is not enough unless students are educated on how to make the necessary changes to their consumption practices to resist such violence.

Of course, offering students to merely take a pledge for extra-credit is not a cure-all or perfect mechanism to educate students on how to implement their beliefs surrounding factory farms and speciesism. At the same time, it causes many students to take the necessary steps to learn how to abstain from consuming animal products. Not all the students in my classes took the pledge. Some students who did take the pledge by the end of the semester decided that they didn't deserve extra-credit because of how short they fell, or because they realized they didn't want a reward for doing what they thought they should have done all along. To this end, one student at the end of the experiment concluded, "I know that one day I will be a vegetarian but at this point in my life I will not despite the clear moral, health, and reasonable positive implications of becoming a vegetarian. I am simply too hedonistic and weak-willed at this stage in my life to become a complete vegetarian." Another wrote,

> sometimes, in refraining from meat, I feel like I can tell myself that I'm making a meaningful, impactful decision, for whatever reasons. Or, I can tell myself that

> ultimately it's really not that big of a deal, or it does not make that much of a difference. This ability 'choose' how I respond to these types of gestures both empowers, but also takes away from the importance of the decision.

One of the most repeated excuses for not wanting to take the pledge was because some students felt that if they were going to go vegetarian they wanted to make the choice for themselves. In fact, students gave me this excuse roughly half the time when they were opting not to take the pledge. Oftentimes, this would occur unprompted after I describe the nature of the pledge to students. Sadly, even though they already admitted that they thought eating factory farmed products was unethical, few students took that moment of reflection as a reason to pledge to themselves to stop participating in the violence irrespective of extra-credit.

Nevertheless, even in these encounters there was an immediate questioning and responsibility placed upon the student in relationship to the course material and how they desired education to take place. At the precise moment the student refused they knowingly opted for their learning to be a disinterested way of earning a grade versus an immediate internalization of what was being taught. As a mechanism of extra-credit students are not punished for deciding not to opt in but they are aware of what they're losing out. Meanwhile students who took the pledge, regardless of their level of success with it, attested to the experience as an enriching way to approach education about non/human animal ethics. One student wrote of the experience, after being unable to eliminate all flesh consumption,

> This failure is not to say that I've completely given up. Now that I am away from home (and its meatly [*sic*] temptations) I have decided to make more of an effort. ... I had a feeling this process would be hard for me, but I have learned that eating meat is a morally disturbing practice and hope I can be determined enough to stop doing it. ... Thank you for bringing the problems of meat consumption to my studies. I don't think I would have really questioned its place in my lifestyle otherwise.

In all cases there was a known choice that each student took, which forced them to critically confront their consumption of course material to either preserve the status quo or challenge it. This then exposed the non-neutrality of their consumption practices. In turn, it became increasingly clear that regardless of what students chose to do in response to grade themselves on internalizing course material and going vegetarian, they all engaged in learning what the material meant in terms of their day-to-day lives.

Ultimately, putting the responsibility on the student for their engagement with the material fosters a non-neutrality that is absolutely critical in getting students to think through the choices they make both inside and outside the ivory tower. While models such as the veg pledge could easily be adopted in a variety of other liberation-focused courses, pledges that focus on the question of consumption puts more on the table than just non/human liberation due to how it's bound up with other avenues of oppression (Calarco, 2003). As Jacques Derrida (1991) writes,

> The question is no longer one of knowing if it is "good" to eat the other or if the other is "good" to eat, nor of knowing which other. One eats him [sic] regardless and lets oneself be eaten by him [*sic*]. ... The moral question is thus not, nor has it ever been: should one eat or not eat, eat this and not that, the living or the nonliving, man or animal, but since *one must eat* in any case and since it is and tastes good to eat, and since there's no other definition of the good (*du bien*), how for goodness sake should one *eat well* (*bien manger*)? And what does this imply? What is eating? ... And in what respect does the formulation of these questions in language...give us still more food for thought? ... The infinitely metonymical question on the subject of "one must eat well" must be nourishing not only for me, for a "self," which, given its limits, would thus eat badly, it must be shared, as you might put it, and not only in language. "One must eat well" does not mean above all taking in and grasping in itself, but learning and giving to eat, learning-to-give-the-other-to eat. One never eats entirely on one's own: this constitutes the rule underlying the statement, "One must eat well." It is a rule offering infinite hospitality. And in all differences, ruptures and wars (one might even say wars of religion), "eating well" is at stake. (pp. 114–115)

To question what one eats is, in essence, an ontological question of what constitutes "the good" and what it means to "eat" in relation to the self and the other. Like with Hooks, Derrida calls upon us to question what consumption is in the first place. Passive uncritical consumption of hegemonic ideologies, institutional practices, and discourses lead us down a path of violence, war, and intolerance. Dis-embodying consumption from its political existence walls off the private from the public in such a way that the status quo is upheld as neutral regardless of what it happens to represent. To eat well means knowing what well is and what it means to consume. What's at stake in offering a veg pledge is far more than a few extra-credit points. At stake is how we consume educational practices and make the classroom more meaningful than just what is said on the page alone.

In the end, for any liberation struggle to succeed it is crucial that academic spaces are taken as the opportunity that they are. To be sure, the classroom is a strong space for testing ideas and role-playing to a certain point. After the debate is over, however, and there is every subjective reason to objectively determine for yourself something is wrong, then there is an imperative to begin to do something about it. If practices don't start to change then the value of such academic safe havens for debate will have been for naught. Supporting racial profiling laws and discriminatory discourse should be incomprehensible to a student who academically agrees in an anti-racist orientation toward the world. Likewise, consuming factory farmed flesh should be incomprehensible to the student with the means who readily decries industrialized non/human agriculture in the classroom. Figuring out ways for students to learn the necessary steps in their private lives, which only they can teach themselves, should be an imperative for any educator looking to foster a truly interactive learning process with an immediate impact. For liberation struggles there is no better time than now for individuals to start withdrawing their acquiescence to oppressive systems of power. While no individual will be perfect, every individual is always able to do more to adhere to their beliefs and make the world better by the way they live their lives. In this sense, a veg pledge model operates as an infinite act of radical vegetarian reflection that asks students to continually withdraw their tolerance where possible for factory farmed flesh, sweat-shop labor, animal-tested products, as well as any other form of consumption they ethically disagree with. In so doing students learn to grade themselves on their own realization of the material. This empowers them to implement their ethics even after the class ends by demonstrating how easy it is to enact the needed agency in the first place. The fact that so many students were able to seamlessly cut out a large percentage of animal products for a handful of extra-credit points proves that it isn't an issue of the change being impossible; but merely an issue on motivating students to take the first step. Opening the door for students in order to show them how easy it can be to challenge institutionalized systems of oppression is precisely the educational initiative necessary to energize a pedagogy of liberation in the academy.

## References

Adams, C.J. (200). *The sexual politics of meat: A feminist-vegetarian critical theory.* New York, NY: Continuum International Publishing Group.

Best, S. (2013). Total revolution: Rethinking radical politics for the 21st century. *TBK2013.* Retrieved from http://www.tbk2013.org/rethinking.html

Calarco, M. (2003). The question of the animal. *EBR.* 2003. Retrieved from http://www.electronicbookreview.com/thread/criticalecologies/animot

CEDA Forums. (2012). Animals, allies, and race. Retrieved from http://www.cedadebate.org/forum/index.php/topic,4217.0.html

Dell'Aversano, C. (2009). The love whose name cannot be spoken: Queering the human-animal bond. *Journal of Critical Animal Studies, 3,* 1–2. Retrieved from http://www.

criticalanimalstudies.org/wp-content/uploads/2009/09/5-JCAS-Vol-VIII-Issue-I-and-II-2010-Ex-Essay-THE-LOVE-WHOSE-NAME-pp-73-125.pdf

Derrida, J. (1991). 'Eating well,' or the calculation of the subject: An interview with Jacques Derrida. In E. Cadava, P. Connor, & J.-L. Nancy (Eds.), *Who comes after the subject?* New York, NY: Routledge.

Gehrke, P. (2010). Being for the other-to-the-other: Justice and communication in Levinasian ethics. *The Review of Communication, 10*(1). Retrieved from http://sc.academia.edu/PatGehrke/Papers/192843/Being_for_the_Other-to-the-Other_Justice_and_Communication_In_Levinasian_Ethics

Guardian. (2010, June 2). UN urges global move to meat and dairy-free diet. *The Guardian.* Retrieved from http://www.theguardian.com/environment/2010/jun/02/un-report-meat-free-diet

Harper, B. (2010). *Sistah vegan: Black female vegans speak on food, identity, health, and Society.* New York, NY: Lantern Books.

hooks, b. (2006). Eating the other: Desire and resistance. In M. G. Durham & D. Kellner (Eds.), *Media and cultural studies: Key works.* Malden, MA: Blackwell Publishing.

Kahn, R. (2009). Marching out from Ultima Thule: Critical counterstories of emancipatory educators working at the intersection of human rights, animal rights, and planetary sustainability. *Canadian Journal of Environmental Education, 14*, 179–193. Retrieved from http://files.eric.ed.gov/fulltext/EJ842748.pdf

Marcus, E. (2005). *Meat market.* Boston, MA: Brio Press.

Marcuse, H. (1965). Repressive tolerance. Retrieved from http://ada.evergreen.edu/~arunc/texts/frankfurt/marcuse/tolerance.pdf

Peterson, S. (2000). Rereading public and private: The dichotomy that is not one. *SAIS Review, 20*(2), 11–29.

Pryse, M. (2000). Trans/Feminist methodology: Bridges to interdisciplinary thinking. *National Women's Studies Association Journal, 12*(2), 105–118.

Regan, T. (2013). Christianity and animal rights: The challenge and promise. *Religion-Online.* Retrieved from http://www.religion-online.org/showarticle.asp?title=2326

Schatz, J. L. (2013). Activist judging: A question of objectivity and pedagogy. *The National Journal of Speech & Debate, 1*(2). Retrieved from http://site.theforensicsfiles.com/NJSD1-2.Final.pdf

Spanos, W. (1993). *The end of education: Toward posthumanism.* Minneapolis, MN: University of Minnesota.

Spiegel, M. (1997). *The dreaded comparison: Human and animal slavery.* London, UK: Mirror Books.

Singer, P. (2001). *Animal liberation.* New York, NY: HarperCollins Books.

Swan, E. (2010). States of white ignorance, and audit masculinity in English Higher Education. *Social Politics: International Studies in Gender, State, and Society, 17*(4), 477–506.

Yancey, G. (2005). Whiteness and the return of the black body. *The Journal of Speculative Philosophy, 19*(4), 215–241.

# 10. *Building Alliances for Nonhuman Animals Using Critical Social Justice Dialogue*

Amber E. George

I am writing this essay within the context of the social justice scholar-activism I have done in Critical Animal Studies (CAS), Eco-ability Studies, and Critical Dialogue. As I write this, conflicts are raging among activists, scholars, and other groups seeking to improve the lives of nonhumans and freeing humans and nonhumans alike from confinement. The conflicts I'm referring to are based on differing perspectives about animal liberation or how to best help nonhumans. Conflicts are often made worse by the complicated nature of identity politics that can generate misunderstanding and prejudice based on one's racial identity, disability status, religion, species, gender, and so on. Inadequately addressing such conflict can hamper nonhuman liberation and also other social justice efforts. Central to diffusing these conflicts is developing one's ability to be an effective ally. Allies are "people who recognize the unearned privilege they receive from society's patterns of injustice and take responsibility for changing these patterns" (Bishop, 2017). This definition suggests an effective ally is reflective, able to communicate well, and take action when it is warranted. However, developing introspection can be a daunting task as many people fall into dualistic or binary thinking that can cause one to avoid carefully examining patterns of injustice and one's beliefs (Perry, 1970). This chapter suggests that critical dialogue can be used as an experiential learning pedagogy that encourages a healthy skepticism toward commonly accepted truisms and nurtures critical thinking. Critical dialogue, in this sense, promotes positives relations among those who share a passion and sense of urgency for justice. The case will be made for how dialogue is a creative tool that practitioners can use to engage estranged groups in conversation about theory and practice from activist, academic, and artistic perspectives. It is especially useful for discussing the intersections of identities, movements, and disciplines to create alliances that pave the way to cooperative coexistence. I will demonstrate how the practices and principles of critical dialogue can serve as a powerful pedagogical framework for the larger field of CAS and all its intersections. I will especially consider how this pedagogy could bring these scholars and activists together to critically explore nonhuman and ecological issues along with other social justice causes. My hope is that the methods garnished from this pedagogy will help us discover new possibilities for mobilization, inclusion, and justice.

For context, CAS is an academic field of study that seeks to abolish animal and ecological exploitation, oppression, and domination. CAS is grounded in a broad global emancipatory inclusionary movement for total liberation and freedom. Best defined, CAS is

> rooted in animal liberation and anarchism, is an intersectional transformative holistic theory-to-action activist led based movement and field of study to unapologetically examine, explain, be in solidarity with, and be part of radical and revolutionary actions, theories, groups and movements for total liberation and to dismantle all systems of domination and oppression, in hopes for a just, equitable, inclusive, and peaceful world. (Institute for Critical Animal Studies, 2016)

Steve Best, cofounder of CAS with Anthony Nocella, argued that previous fields such as human/animal studies have been appropriated by scholars who prioritize career over effective advocacy, incorporation over activism and confrontation (2007). Like other forms of anti-oppression scholarship and activism, CAS is a movement that politicizes the experiences of nonhuman animals with the hopes of securing support from new members seeking liberation for all. Best (2007) argues that the "stultifying structure of 'higher education' defuses the potential volatility of critical knowledges." Most especially those that may "expose the true horror of the animal holocaust and international animal slave trade, which exploits, tortures, and murders burgeoning billions of victims." Furthermore, Best (2007) states the academic-industrial complex itself, in its highly profitable vivisection sectors, butchers over one hundred million animals a year for "medical research." However, contrary to this contention, I argue critical social justice pedagogy could bring together scholars and activists, academia, and community to entice a new generation of educators, thinkers, and activists to join in the fight for anti-speciesist praxis. Dialogue has the promise of bridging the tension often observed between theory and practice, community and campus, scholarship and citizenship. Furthermore, the foundation of critical social justice dialogue sets allyship as the best way to promote systemic change and avoid stuffy elitism, abuse, and neutrality that Best says is rampant in nonhuman realities and traditional animal rights campaigns (2007).

The type of pedagogical dialogue I'm advocating is influenced by the Intergroup Relations (IGR) program that originated nearly 25 years ago at the University of Michigan. The program was originally created to address racial tensions; however, it has gained international acclaim for successfully raising awareness, coalition building, and generating social change related to issues of racism, sexism, heterosexism, ableism, and other forms of oppression (Michigan, 2016). The Michigan model offers insight for how animal activists embroiled in conflict can use small-scale dialogic communication processes, whether online or in person, to help communicate their positions and consider the positions of others. This pedagogy emphasizes learning, listening, and developing shared understanding as a community-based praxis (Cissna & Anderson, 1994). Intergroup dialogue, as outlined in the Michigan model, is defined as "face-to-face facilitated conversations between members of two or more social identity groups that strives to create new levels of understanding, relating, and action" (Zuniga, 2003). Social identity is part of an individual's sense of self as derived from perceived or actual membership in a social group, and is acquired or proscribed. According to Zuniga, examples of groups that have participated in intergroup dialogues include men and women, white people, biracial/multiracial people, and LGBTQ and heterosexual people, among other social identity groups (2003). Over the last two decades, educators, social workers, community leaders, activists, social scientists, and other practitioners have used Michigan's model of critical dialogue to build relationships through the power of transformative experiences that foster deep reflection and enhanced understanding of differing perspectives (Judkins, 2012).

## Fractions within CAS and Nonhuman Animal Justice Movements

Often social justice endeavors seek to promote feelings of unity, acceptance, and tolerance within a given culture (Wilson, 2014). However, generating tolerance, for instance, is more likely to lead to critical erasures of perspectives along with setting up unequal relationships and, therefore, is not enough. One of the more recent critiques of animal rights movements is that it is a largely white, heterosexual, upper-class, able-bodied and able-minded, privileged movement. While there are plenty of people within animal justice movements that identify as feminists, allies with People of Color, and rally against all forms of bigotry, there remains a gaping hole in promoting inclusion for all and upholding many of CAS' principles. For instance, racial segregation in animal rights movements is a problem according to the vice president of the U.S. Humane Society, who has suggested that "less than three percent" People of Color participate (Hamanaka & Basile, 2005). When People of Color do enter CAS conversations or activism, they are oft met with insensitivity, tokenism, prejudice, and an overall unwelcoming environment (Harper, 2010; Wolbring, 2014). For instance, Kris, a Black activist, recounts her experience of tokenism, "They haven't done outreach to the community, but they call—'Hey we need a Black face at the protest.' I go, but it's not a unifying way, it's a marginalizing way of organizing. You're not one of us, but we need you" (Hamanaka & Basile, 2005). Whereas in other segments of the nonhuman animal activist movement, ableist terms are often used to denote negative aspects related to nonhuman oppression. Such is the case with the well-known animal scholar-activist Gary Francione's coinage of "moral schizophrenia" to describe those who care for animals such as cats but exploit other animals such as cows for food, clothes, and entertainment (2000). Francione used the term schizophrenia in a negative manner, to stigmatize those who condone animal abuse, and in doing so insults people who identify as having schizophrenia. Moreover, who can forget Princeton philosopher, Peter Singer, an advocate of animal rights, who supports euthanizing infants with disabilities by unilaterally denying them access to life-sustaining health care? Furthermore, PETA has received negative press for their commodification of women and sexualizing animal activism in their campaigns against animal abuse (Adams, 2013). Moreover, the rampant bullying of vegan activists on social media forums recently inspired the creation of a Facebook online support group called *In Defense of Animals*. People can join the group to discuss the real ramifications of activism including such issues as burnout, self-care, bullying, compassion fatigue, anger, guilt, family issues, activist infighting, and more (In Defense of Animals, 2016). The fact that such a group exists suggests there is mounting tension among individuals and groups seeking to promote veganism, be responsible stewards of the land, and advocate for nonhumans. The conflict is often the result of humans not recognizing their privileged identities, or knowing but not acknowledging how their interactions can influence individuals who have differing experiences.

CAS's commitment to inclusion is a prominent part of its vision, just the same as Michigan's Intergroup Relations is committed to fostering connections in academia. One of the guiding principles for creating critical dialogue among nonhuman activists is ensuring an abundance of different perspectives are present in dialogic conversations. In these spaces, attention is drawn to those who are present and who might be absent, so far as social identity is concerned. Then, the visibility or invisibility regarding the social and structural conditions that foster social conflict are discussed. Despite the richness of diversity of those involved in CAS, without the tools to communicate across various backgrounds and experiences, participants may feel isolated and require help with bridging these gaps. These communication tools involve asking questions, building self-awareness, being honest and flexible, listening actively, avoiding stereotypical thinking, and respecting difference, just to name a few. There is a deliberate attempt to create an open and safe space in which participants can explore what is often omitted or invisible, by highlighting the importance of interdependence, partnership, and relationships.

Many trace the epistemological underpinnings of dialogue for transformative praxis back to Freire who suggests we develop an awareness of the ways in which mutual learning can create meaningful change (1970). Participants share existential experiences, developed through deep immersion in existing social problems while learning how to manage difference and conflict. Social reality is explored as a phenomenon that links subjective experiences to social experiences. In other words, while surface readings of reality are essential, it is the depth of critical consciousness that will enable one to discover a new reality through the process of dialogic inquiry. Consciousness raising is another aspect Freire (1970) emphasized as it allows members of oppressed groups to understand the historical circumstances behind their oppression. Privilege stemming from interpersonal, institutional, and societal power dynamics also influence these interactions. It requires critiquing the social relations, institutions, and traditions that create and maintain systems of privilege and oppression. Thus, individuals participating in the dialogue groups explore these conditions and go beyond purely theoretical comprehension of the issues. Rather, they generate meaning together by exploring their personal realities through a dialectic between reflection and action. The meaning that comes from doing this work together tends to inspire one to partake in action and is absolutely necessary for social transformation, or as Freire (1970) defined, "educational praxis."

Another important theme in Freire's work is his focus on literacy communication; both literally in the ability to read and write, and figuratively through the power and confidence that knowledge instills in the learner. Dialogue is ideal for producing knowledge through communication because its "preeminent purpose is to conscientize the people" (Freire, 1970, p. 47). When discussing CAS principles, three modes of communication—debate, discussion, and dialogue—may be used; however, dialogue is preferred. Citing my own experiences, I have observed that many animal studies and

CAS scholars engage in debate style forums to discuss critical CAS issues. This is useful when evidence-based thinking is needed, or one seeks to enhance presentation skills and strengthen one's ability to influence others by defending one's position (Keller, Whittaker, & Burke, 2001). However, debate tends to create situations where one is being evaluated, judged, and critiqued for their viewpoints and feelings. Instead of focusing on the idea or opinion being expressed, people get attacked which can cause participants to disengage from further discussion. Unlike debate and some forms of discussion, dialogue fosters empathic connection, understanding through inquiry, and mutual respect (Bohm, 1996). Sincerely empathetic people suspend judgment and criticism when they are listening, without adjudicating one's feelings or actions. Used here as part of reflecting on and engaging in CAS, dialogue is a medium for sharing empathy—or the ability to experience another's feelings whether joy or pain. Participants build an understanding of, and empathy for, the narratives of others, including nonhumans. Empathy, leading to greater understanding and the ability to communicate about differences, is achieved regarding the opinions expressed and the underlying needs, assumptions, values, and concerns (Bohm, 1996). In the process, participants build a clearer understanding of their perspectives and others' viewpoints. In many cases, one's perception shifts, and we move from simply staring at the objective, static world of things to relishing in the story of the living being, or a living system and, of course, connecting more deeply with the self.

One method for connecting empathetically with each other is through storytelling. While storytelling can bring communities together in solidarity to create an inclusive, larger narrative that builds on hope and positivity, it can also create division. Division may arise if one doesn't listen with the intent of fully understanding, or pretends to be listening when really they are simply waiting for their turn to share. Furthermore, all members must feel comfortable expressing themselves and participating fully, without fear of retaliation, erasure, judgment, or disavowal of experience. Initiating this within scholar-activist spaces is revolutionary because the norm is to engage in debate or discussion that often denies the personal, emotional, raw experiences of all to favor the rational, practical voices of privileged humans. Building on a decolonial position, we must reject the Western, or more precisely Euro-American, beliefs that winning arguments and proving "truth" through patriarchal and individually oriented theory and fierce debate is an effective long-term process for social change. We must maintain that storytelling and open free-flowing dialogue, rather than just oral and written argumentation, is a viable avenue for communication.

Another unique aspect of this pedagogical method is that participants are not "taught," rather each participant shares a role as learner and teacher. This arrangement helps dismantle the usual power hierarchy of one, all-knowing teacher responsible for filling the (read: empty) minds of many unknowing students that Freire (1970) discusses. In this setting, education is nurtured via facilitators who have been trained by facilitation pedagogy, exhibit strong leadership skills, and exude passion for social justice work. These facilitators are not the authority figures or "teachers" of the dialogue, rather everyone has an equal part in the learning and teaching. This pedagogical strategy is known as colearning and can be summarized as follows: it is a mutual learning process exemplified by the phrase, "I am one of the teachers and one of the learners in this experience." It is collaborative; a community of cooperation is created whereby conversations take shape with centers not sides, and move away from competition and exclusion toward increased partnership. Thus, the ideology of participatory, inclusive education, the early stages of which were discussed by Freire (1970), is particularly important. It is structured to create a self-governing process that acknowledges and respects all parties. The primary philosophy behind this movement is mutually subjective and respects others just as CAS praxis does within the nonhuman animal activist movement.

The pedagogy for facilitating discussions among conflicted groups involves a range of exercises designed to build awareness of social identity and enhance group dynamics, communication techniques, and empathy. One such exercise I've used in my dialogue groups is what I call the Social Identity Pie. The exercise requires students to self-describe their social identities according to their race, gender, sexual orientation, class, ethnicity, socioeconomic status, and disability status. Then participants create a pie chart that designates a slice of pie to represent how aware they are of each identity on a daily basis. The more aware they are of the identity, the bigger the slice. It is often the case that one's privileged identity, such as heterosexuality or whiteness, will receive a smaller slice whereas one's non-privileged identity such as LGBTQ or disability status may be drawn larger. Next students are paired up to engage in a speak and listen activity whereby each person takes a turn explaining their pie to another person who simply listens and affirms. Participants also respond to some prompts such as "which identities are you more aware of? Which identities do you not think about often? Why do you think that is?" Once the person is finished speaking after two minutes, then the listener must repeat back what the speaker said about their pie and responses to the prompts. Participants switch off roles until everyone has had a chance speaking and listening. At the end of the exercise, participants gather into a large circle to have a dialogue using debriefing questions such as "how did it feel to do this exercise? How does this exercise align with the values of CAS? How might this information help you to be a better ally?" The results of this exercise are plentiful and include raising awareness of self and other, pushing one's comfort zone and admitting vulnerability, practicing dialogue skills of listening and speaking, and generating ideas for becoming active allies for the self and others within CAS and social justice movements.

Within this pedagogical strategy, conflict often lies at the heart of interactions. I have witnessed participants feel conflicted during the debriefing of the Social Identity Pie exercise mentioned above. Conflicts, such as the ones between those who support total liberation and those who support welfare reform for nonhumans, are valuable opportunities for participants to engage in heart-to-heart conversations. Then within this conflict often lies subliminal conflicts among participants based on social identity, privilege, and non-privilege. Conflict of opinion, viewpoint, and understanding is a natural and even beneficial part of the dialogue process. It will take place in various ways within the group—within individuals, between individuals, or between groups. It may be overt or covert in the group, and present, but not fully recognized. Unlike "feel-good" types of encounters that attempt to promote understanding by avoiding, masking, or "overcoming" social injustices, critical dialogue recognizes that communicating about and working through difficult feelings and injustice is a necessary part of the encounter. The impassioned responses of those scholars and activists who believe their viewpoint is correct, and sometimes only valid perspective, can be explored within such a context if all parties are open to doing so. In particular, empathetic listening is crucial to dealing with conflict, as it can generate a sense of catharsis. The connection among participants occurs when the person sharing feels that they are among empathetic listeners and are being taken seriously. In attempting to manage disputes, the goal is to be less vulnerable to emotional triggers that may spark continued misunderstandings and may even spur violence. Thus, one of the main tasks in dialogue is to transform conflict and discomfort into learning for everyone.

I have learned that participants seem more likely to work through conflict when they feel that the environment is safe and the people present are committed to learning. Thus, dialogue circles need to create a safe, yet brave space where comfort zones can be expanded to spawn learning. To this end, specific ground rules are created before engaging in dialogue, which are unique to the needs of the group and created through democratic processes. They are hung on newsprint paper during every meeting. Examples of some ground rules I have been privy to in CAS circles include

1. One person communicates at a time, while everyone listens, without interrupting.
2. Communicate only for yourself, not as a representative of any group.
3. Respect the privacy of what is shared in dialogue. Do not tell other's stories outside of dialogue unless granted permission.
4. No cell phones out during dialogue time.
5. Communicate truthfully.
6. Share your discomfort. If something said bothers you, say so and why.

Social justice learning environments are known as being spaces defined as "safe." However, much speculation has gone into understanding what "safe" means and questioning whether "safe" spaces are what is required to achieve justice. A collective vision for "safe space" is one in which participants feel comfortable expressing themselves and participating fully, without fear of retaliation, judgment, or violence. Brian Arao and Kristi Clemmens had this to say about safe spaces:

> Further, we share the conviction that violence of any kind—physical, emotional, and psychological—is antithetical to the aims of social justice work; indeed, we see the use of violence to achieve one's goals as a patriarchal norm that should be challenged through such work. (2013, p. 139)

Social justice educators should take the safe space one step further by incorporating a notion of "brave space." Creating a safe, yet brave space means being willing to accept that within any social justice learning environment, the discomforts of discussing one's power and privilege are likely to initiate hostility and resistance. To foster critical, yet real conversations that are likely to initiate social change, one must be daring enough to be vulnerable because, "learning necessarily involves not merely risk, but the pain of giving up a former condition in favor of a new way of seeing things" (Boostrom, 1998). When having a dialogue with those who disagree with us, wearing a badge of courage instead of a shield of safety allows one to air their grievances when it comes to discussing controversial topics like, for instance, keeping nonhumans in captivity.

One of the major ways in which we generate brave yet safe places is by employing the use of biographical testimonials. Sharing these personal narratives about a topic, such as one's decision to become vegan, for instance, assists participants with building trust and interpersonal connections. As such, each participant is required to share an 8–10 minute reflection (often constructed prior to class, however many do it improvisationally) which deals with a prompt. In the case of a dialogue between animal activists who want to discuss how racial tension affects one's decision to adopt veganism, for example, having them share a deeply personal story about their race and veganism may help them bridge differences, find common ground, respect contrary viewpoints, yet ultimately bond as allies committed to animal justice or CAS principles. During a session of testimonial sharing, all participants sit in a circle facing each other. Each participant takes a turn sharing their story during which all other participants listen attentively. No one speaks until after everyone has shared their testimony. Facilitators and participants discuss how they feel at the moment, what surprised them, and are given the opportunity to ask other participants questions about their stories. Facilitators may ask questions such as, "what kinds of feelings came up for you when you heard the story or experience different from your own?" The outcome of such a personal experience is that participants often appreciate learning about their differences and similarities. Participants also engage in critical self-reflection to appraise one's personal perspectives, prejudices, and privilege. Engaging one's self through the sharing of stories encourages participants to take risks and admit vulnerabilities. The exercise can also generate feelings of moral indignation, which if used positively can transform one into being more proactive. In this brave, yet safe space, participants move away from independence, competition, and exclusion to increased partnership and inclusion.

## *Critical Social Justice Dialogue Implications for ICAS Praxis*

The process of building bridges can provide participants with a sense of shared responsibility for challenging oppressions and feeling of accomplishment when change is met; however, with these successes, comes its own batch of struggles. Since engaging in social justice dialogue is a process by which change occurs slowly through relational change rather than structural change, some people can find it challenging to address larger structural issues in an immediate way. For those who join in the struggle, put forth the effort, and do not see immediate and profound societal change, the whole process can be frustrating. To avoid individuals abandoning the cause or giving up, participants must initiate action outside of the dialogues. For deep, sustainable transformational change to occur, it must be linked to action at the personal, interpersonal, intra- and intergroup levels. For those in animal activism and doing CAS work, action may take the form of protests, writing articles, blogging, spreading awareness through social media, and other opportunities. One of the final goals of critical social justice dialogue then is for participants to put into action the skills and commitments they developed in the dialogue to settings outside and beyond the dialogue.

Once participants leave the dialogue space and engage with the issues on-the-ground, an immediate need for camaraderie, support, and encouragement may arise. Thus, the pedagogy of critical dialogue can promote self-care and the care of others. Far too often those engaged in these struggles cannot find the support networks they need to deal with overwhelming feelings of pessimism, anxiety, guilt, isolation, irritability, anger, and fatigue. It is also beneficial to have a support network to share in the successes and joys of eliciting transformation. One way that critical dialogue circles can provide support is by listening to each other's stories of success and failure, lift each other up when feeling down, and remind each other to take care of themselves. With all that is wrong in the world, asking highly passionate activists to slow down, "take care of yourself" and "be well" is hard. There are times when we feel otherwise unmotivated by the same activist struggles, events, or discussions that used to give us strength. This is when we know we've stretched ourselves too thin and may be burnt out.

Sometimes activists can get cranky when you ask them, "are you taking care of you?" It's almost like when we admit to a little self-indulgence, or of "taking the day or weekend off" it is a privilege or weakness that suggests we aren't committed enough to the cause. Competing forces such as economic struggles, family obligations, workplace expectations, or even the very structures of the movements can intentionally and unintentionally work against one's efforts to take care. Activists should not be quick to accuse ourselves, and one another of not being radical enough, committed enough, or tough enough. As scholar-activist, Audre Lorde said, "caring for myself is not self-indulgence, it is self-preservation, and that is an act of political warfare" (1988, 132). If we are going to foster a truly inclusive society, we must do the systemic work of making wellness accessible for all, without judgment or setting unrealistic expectations. This may be the biggest, yet the simplest act of inclusion we can undertake.

However, adopting an ethos of self-care in an activist's world doesn't just happen. Saying you need to take care to avoid burnout or overcommitment and knowing you need to do it does not make it happen. As scholar-activists, making it happen means changing habits and schedules, and shifting deeply held beliefs about self-worth and identity. Retail therapy, junk food, and suppressing one's emotions are very unlikely to provide nourishment and generate spirit. A good bubble bath (with products not tested on nonhumans!) can take you little ways in this direction, but long-term self-care is not about luxury and consumerism. Nor is it about escape. It is about finding those things that feed and sustain you. While this could be different for everyone, I can state with clarity that for me, it has meant finding opportunities to connect with others, nonhuman and human alike, to find joy, gratitude, love, and a sense of purpose. In the language of ecofeminist and philosopher Joanna Macy, these things have helped me "come back to life" when I have felt very discouraged (Macy, Brown, & Fox, 2014).

When feelings of discouragement or burnout rear their ugly heads, I pay heed to Maya Angelou's advice, "Do the best you can until you know better. Then when you know better, do better" (Egan, 2015, p. 19). Or perhaps we might suggest, "Do the best you can, until you can do more. Then when you can do more, do more." Developing new habits of self-care requires prioritizing, honest reflecting, and asking your friends for help. The point is that none of us learns or can make a difference in isolation; we are interdependent and must engage in a little give and take. Overall, our scholar activism is improved by generating dialogue and practicing care because if we are going to create an anti-oppressive world, we must uphold the dignity of all. This includes not only the beings we are supporting but also other scholar-activists and their causes.

Inevitably, scholar-activists are often on the receiving end of significant pushback from those in academia who disagree, feel threatened, or seek to fuel your demise. Every day you will be surrounded by people and institutions that will reward and punish you. It seems that many individuals in the academy continue to downplay the work of scholar-activists as it is those who study a topic, research conclusions, and place their results in a journal without doing anything else about it are often held in higher regard. This strikes me as troubling and even threatening to the future of social justice, but unfortunately, it makes sense. As outlined above, it is far more difficult to be a scholar-activist, as it takes time, energy, emotional labor, and a thick skin. Not to mention, it is most often an unpaid gig. However, if you adopt a public agenda of scholarly activism, you can bring the lessons of scholarship into the real lives of communities. It will challenge you to become clearer about your ideas, more thoughtful about how to communicate them, and wiser about the merits of research. When you're struggling to fill your roles as scholar and activist, make your passion for justice not only your mission but its own reward. Choose to work with people that lift you up, inhabit places and spaces that can provide the support that a genuine scholar-activist requires. Cultivate relationships with other scholar-activist comrades who share your activist commitments and can come to your defense if the occasion arises.

In conclusion, the key principles, practices, and objectives that underlie critical dialogue and CAS support are raising the consciousness of all, building relationships across differences, and strengthening individual and collective skills for social justice. I have witnessed first-hand how the practice of critical dialogue is a substantive and meaningful avenue for preparing individuals with the knowledge, commitment, and skills essential for living and working in a diverse society. Participants gain an understanding of their social identities, and the effects it can have on social inequity at personal, interpersonal, and structural levels. Additionally, participants develop communication, leadership, and alliance building skills to work in various scholarly and activist roles. Students leave the process with individual and collective capacities to address social justice issues on their respective campuses and beyond. However, this type of work can be exhausting, and progression toward the goal can be slow moving. When you feel like giving up because you are burned out and tired, remember why you began this work in the first place. Remember that we are activists because our role gives us satisfaction, and in many cases is imminent to our survival. Recall that when you fail, you were working to make our society fairer, more just, and kinder. Remember this role can give us friends like no other. And remember our activist efforts illuminate our world in ways that scholarship alone never can.

## References

Arao, B., & Clemmens, K. (2013). From safe to brave spaces: A new way to frame dialogue around diversity and social justice. *The Art of Effective Facilitation.* Retrieved from https://sty.presswarehouse.com/sites/stylus/resrcs/chapters/1579229743_otherchap.pdf stylist publishing.

Best, S. (2007). The rise (and fall) of critical animal studies. Retrieved from http://www.liberazioni.org/articoli/BestS-TheRise%28and%20Fall%29ofCriticalAnimalStudies.pdf

Bishop, A. (2017). Becoming an ally. Retrieved from http://www.becominganally.ca/Becoming_an_Ally/Home.html

Bohm, D. (1996). *On dialogue.* New York, NY: Routledge.

Boostrom, R. (1998). Safe spaces: Reflections on the educational metaphor. *Journal of Curriculum Studies, 30*(4), 397–408.

Cissna, K. N., & Anderson, R. (1994). Communication and the ground of dialogue. In R. Anderson, K. N. Cissna, & R. C. Arnett (Eds.), *The reach of dialogue: Confirmation, voice, and community.* Cresskill, NJ: Hampton Press.

Egan, J. (2015). *3,000 Astounding quotes.* Retrieved from https://books.google.com/books

Francione, G. L. (2000). *Introduction to animal rights: Your child or the dog?* Philadelphia, PA: Temple University Press.

Freire, P. (1970). *Pedagogy of the oppressed.* New York, NY: Continuum.

Hamanaka, S., & Basile, T. (2005). Racism and the animal rights movement. *Satya Magazine.* Retrieved from http://www.satyamag.com/jun05/hamanaka.html

Harper, B. (2010). Race as a 'feeble matter' in veganism: Interrogating whiteness, geopolitical privilege, and consumption philosophy of 'cruelty-free' products. *Journal for Critical Animal Studies*, *8*(3), 5–27.

*In Defense of Animals*. (2016, March 20). Let's tackle vegan bullying together. Retrieved from https://www.facebook.com/indefenseofanimals/photos/a.10150796907537346.439113.5956327345/10153980355292346/?type=3

Institute for Critical Animal Studies. (2016). Retrieved from http://www.criticalanimalstudies.org/

Johnson, L., & Thomas, S. (2013). Interview with Carol J. Adams. *Journal for Critical Animal Studies*, *11*(1), 108–132.

Judkins, B. M. (2012) Intergroup dialogues, building community and relational justice. *Catalyst: A social justice forum*, *2*(1), 6.

Keller, T. E., Whittaker, J. K., & Burke, T. K. (2001). Student debates in policy courses: Promoting policy practice skills and knowledge through active learning. *Journal of Social Work Education*, *37*(2), 343–355.

Lorde, A. (1988). *A burst of light: Essays*. Ithaca, NY: Firebrand Books.

Macy, J., Brown, M. Y., & Fox, M. (2014). *Coming back to life: The updated guide to the work that reconnects*. Gabriola Island, Canada: New Society Publishers. Perry, W. G., Jr. (1970). *Forms of intellectual and ethical development in the college years: A Scheme*. New York, NY: Holt, Rinehart, and Winston.

Program on Intergroup Relations. (2015). IGR History. Retrieved from https://igr.umich.edu/article/igr-history

Wilson, H. (2014). The possibilities of tolerance: Intercultural dialogue in a multicultural Europe. *Environment and Planning D: Society and Space*, *32*, 852–868.

Wolbring, G. (2014). Ability privilege: A needed addition to privilege studies. *Journal for Critical Animal Studies*, *12*(2), 118–141.

Zuniga, X. (2003). Bridging differences through dialogue. *About Campus*, 9. Retrieved from http://people.umass.edu/educ202-xzuniga/downloads/ZunigaAboutcampus.pdf

# *Part VI: Decolonizing*

# 11. Decolonizing Critical Classrooms, Decentering Animal Pedagogies

William E. Shanahan III

> Critical Animal Studies should be both a field of study and a movement focused on telling and fighting for the truth about the treatment of other animals by many humans, and this truth clearly threatens the powers that be. (Nocella, White, & Cudworth, 2015, p. 8)

study field—
moving, telling, fighting
for
truth, thankfully
not
Truth

> In war, truth is the first casualty. (Aeschylus, n.d.)

in war,
truthfully, the first
casualty is never
truth

> History, in other words, belongs to the victors. Perhaps no one understands this better than indigenous peoples who, in addition to suffering the depredations of genocide, colonization, and cultural annihilation, have been revictimized at the hands of whitestream history. The lesson here is pedagogical. (Grande, 2004, p. 242)

genocidal, colonial, cultural
depredations:
preying, plundering,
robbing, ravaging, re-

casting

> Through "colonialism" was reshaped the material basis of exchange that determined human relationships. Put differently, the very idea of what is "human" was recast by the imposed value-systems of the "civilizing" process that was colonialism To be human, to live, and to relate to others, thus, both lost and gained meaning. (Nayar, 2002, p. 111)

white-
streaming pedagogical lessons
here, re-
casting civilizing processes
then, im-
posing human be-
ing (there)

> Total liberation is greatly influenced by anarchism in that it is opposed to all forms of oppression and domination and is also not reformist. (Nocella et al., 2015, p. 13)

opposing all
oppression, domination
except
opposing all
oppression, domination

> Critical Animal Studies grounded in anarchism is intersectional and radical. It stands against all systems of oppression and domination, and promotes activism and community organizing. (Nocella et al., 2015, pp. 7–8)

standing, promoting, organizing:
activating against, for
total liberation

> I say that absolute liberty is an abstraction not found in this world but that conditional liberty is possible, desirable, and should be sought. ...Our minds and bodies are restricted in many ways. Nor does our education help us to reduce these restrictions. From our very childhood we are subjected to a process of socialization and enculturation (to use ugly words for an ugly procedure) compared with which the training of household pets, circus animals, police dogs is mere child's play. (Feyerabend, 1982, p. 174)

except opposing all
socializing, enculturating,
opposing all

household pets
circus animals
police dogs,
opposing all
ugly
training

> Anarchism is a socio-political theory which opposes all systems of domination and oppression such as racism, ableism, sexism, anti–LGBTTQIA, ageism, sizeism, government, competition, capitalism, colonialism, imperialism and punitive justice, and promotes direct democracy, collaboration, interdependency, mutual aid, diversity, peace, transformative justice and equity. (Nocella et al., 2015, p. 7)

except opposing all
domination, oppression
systems,
except opposing
promoting peace,
punishing states

> So elementary education joins hands with higher education to produce individuals who are extremely limited, unfree in their perspective though not at all in their determination to impose limitations on others under the name of knowledge. (Feyerabend, 1982, p. 175)

opposition rubs
promotion:
knowing opposing
risks re-
casting
civilization,
colonialization, en-
culturation

> The imperative before us, as educators, is to ensure that we engage a thorough examination of the causes and effects of all wars, conflicts, and inter/intracultural encounters. We must engage the best of our creative and critical capacities to discern the path of social justice and then follow it. The ongoing injustices of the world call educators-as-students-as-activists to work together—to be in solidarity as we work to change the history of empire and struggle in the common project of decolonization. (Grande, 2004, p. 242)

our privilege, all

our privilege:
here, others not
here—
ancient forests, massive
oaks, chestnuts, hickories, walnuts, beech, giant
hemlocks, soaring one hundred feet, towering
canopies: chopped, burned, pulped,
gone—
query passenger
pigeons, once
billions, now
none,
eastern elk and cougar, sea mink,
bats, bears, bobcats,
now, none: our
privilege here,
others cannot
(mishawum, mystic, massachusett,
cowate at charles river falls,
descend across ten thousand
years, en-
slaved, sold, executed, infected,
extinct-
ed)

> A successful rethinking of world politics…must deal with issues of representation. It must engage the languages through which we have come to distinguish the safe from the threatening, the rational from the irrational, the possible from the impossible. Needed is a critique of language that opens up possibilities to gaze beyond the givenness of world politics, that can problematize political dilemmas that have been rendered unproblematic, even invisible, through years of normalizing speech and corresponding political practices. (Bleiker, 2000, p. 271)

safe, threatening
rational, irrational,
possible, impossible:
language engages,
engages language,
"gaze beyond the givenness,"
"problematize political dilemmas"
critique language,
rethink politics,

open possible—
representation issues challenges,
forms substance,
aestheticizes politics,
collapses separation,
not neutral,
not transparent,
not detached:
attach social, cultural,
transform political,
obstruct transform,
obstruct enable,
enable obstruct

> The difficulties the Zapatistas had in explaining their notion of revolution to the Indians were of course compounded by the seemingly simple barrier of language. ...It became clear that the indigenous held a different and, to the non-Indian, a curious conception of time. ...The mestizo Zapatistas realized that their problems were not simply problems of translation. They became aware that the Indian language had its own, different, referents, its own cultural markers. (Higgins, 2000, pp. 363–364)

curious
conceptions of time,
mis-
understandings, in-
comprehension:
now,
five hundred years ago,
when time began—
culture, perspective, education,
irrelevant:
politics precede, sub-
sume, over-
ride, ride over
culture, perspective,
educate, contaminate, coordinate:
referents mark culture,
language refers difference,
in dios owns, refers, marks,
differs

> Ultimately, the essay [by Richard J. White] argues that strategies focused on total liberation, which are sensitive to the interconnected oppression and violence affecting human and other animals, need also to pay attention to the imperative to liberate "the spatial" landscapes, and disturb the normalization of violence toward nonhuman animals. (Nocella et al., 2015, p. 18)

normalizing violence
toward
liberating landscapes
disturbing
normalizing violence
toward

> Every protocol had been observed between the whaling chief and the spirit of the whale. Keesta had thrown the harpoon, and the whale had accepted it, had grabbed and held onto the harpoon according to the agreement they had made through prayers and petitions. Harmony prevailed, whaler and whale were one, *heshook-ish tsawalk*. (Atleo, 2013, p. x)

promoting direct
democracy,
collaboration, interdependency, mutual aid,
diversity,
peace

> Ah-up-wha-eek (Wren) landed on the whale and spoke to Keesta: "Tell the whale to go back to where it was harpooned." Keesta spoke to the whale, and immediately the great whale turned according to the word of Wren, the little brown bird, and returned to where it was first harpooned, and there it died. (Atleo, 2013, p. x)

dis-
harmony, dis-
unity:
ritual, prayers, petitions

> The politics of place in Native American studies is very tricky both socially and politically. While conceptions of Native identity are legislated differently depending on governing nation-states, tribal government systems, histories, and cultural differences, they share spatialized tendencies, identity, social relations, and politics are often conceived, represented, and determined as geographically and historically situated and bound to a particular community. (Goeman, 2009, pp. 169–170)

opposing all settler-
colonial
spatial abjecting
legislative slaughtering
ripping (out) grounding

> This grounding, even while considered abject space by the settler state, is of utmost importance to the imaginative geographies that create the material consequences of everyday existence for Native people, even while the historical onslaught of legislation continues to rip that grounding out from under Native people. (Goeman, 2009, p. 170)

imagining, con-
sequencing existing,
grounding everyday,
creating material—
imagining geographies, re-
casting geography

> Opening up thinking space involves both opportunities and challenges. The task is clear: to find new ways of thinking, seeing, hearing and sensing the political; to break free of disciplinary boundaries; to write creatively and to explore other ways of communicating. (Beiker, 2017, "Navigating," para. 1)

opening spaces—
thinking, seeing, hearing, sensing,
imaginative, creative
spaces:
closing,
disciplining,
ripping,
bound-
ing
historical
material
geographical
political
legislative
grounding,
spacing,
communicating opening
up

> A path must be cleared for a new interpretation. ...The new hierarchy does not simply wish to reverse matters within the old structural order. ...A new hierarchy and new valuation mean that the ordering *structure* must be changed. (Heidegger, 1979, pp. 209–210)

reversing
breaking free
of
geographical grounding
disciplinary bounding
structural ordering
hierarchical valuing:
reversing
of

> Only then is Platonism overcome, which is to say, inverted in such a way that philosophical thinking twists free of it. (Heidegger, 1979, p. 201)

Herausdrehung:
invert,
twist free, over-
come, dis-
place, de-
construct
breaking free

> But then where does such thinking wind up? During the time the overturning of Platonism became for Nietzsche a twisting free of it, madness befell him. (Heidegger, 1979, pp. 201–202)

winds up
with in
sanity, in-
sanity

> Language is an obvious starting point. Existing modes of knowing are inevitably linked to existing ways of speaking and writing. To think differently is to stretch the boundaries of language. (Bleiker, 2017, "Navigating," para. 2)

stretch, not
break,
boundaries—
knowing links
speaking

writing, or
writing
speaking links
knowing, or
knowing be-
speaks
writing: perhaps
invert, reverse, overturn
twists free from, overcomes
stretching
breaking
liberating

> For the history of Western thinking in the past eight centuries has been one of replacement of ideas within a framework that has remained basically unchanged for nearly two millennia. Challenging this framework of interpretation means a rearrangement of our manner of perceiving the world, and it involves a reexamination of the body of human knowledge and its structural reconstruction into a new format. (Deloria, 1999, p. 101)

challenging everyday
existing, re-
arranging, re-
examining, re-
placing, re-
constructing
perceiving
interpreting
knowing, be-
ing

> Example after example could be cited, each testifying to the devastating effect of a general attitude toward the world that underlies the Western approach to knowledge. The basis of this attitude is the assumption that the world operates in certain predetermined ways, that it operates continuously under certain natural laws, and that the nature of every species is homogenous. (Deloria, 1999, pp. 102–103)

minority scholarships,
low-income housing,
bilingual education:

> We assume we know the structure of reality and must only make certain minor adjustments in the machinery that operates it in order to bring our institutions into line. (Deloria, 1999, p. 101)

machinery structures
reality,
adjust machinery
adjust structures
adjust reality—
just minor, minor
justice

> If we are then to talk seriously about the necessity of liberation, we are talking about the destruction of the whole complex of Western theories of knowledge. (Deloria, 1999, p. 106)

destroy complex
destroy theories
destroy knowledge,
liberate
machinery
structures
reality

> If we change the very way that Western peoples think, the way they collect data, which data they gather, and how they arrange that information, then we are speaking of liberation. (Deloria, 1999, p. 106)

change data
collecting
gathering
arranging:
adjust
change
destroy
thinking

> For it is the manner in which people conceive reality that motivates them to behave in certain ways, that provides them with a system of values, and that enables them to justify their activities. (Deloria, 1999, p. 106)

reality motivates
behaving
valuing

justifying,—
serious talk
necessary liberation
destroy reality

> Space is not merely a backdrop, though, against which the communication of cultural politics occurs. Rather, it needs to be recognized as a central component in that communication. It functions as a technology—a means and a medium—of power that is socially constituted through material relations that enable the communication of specific politics. (Shome, 2003, p. 40)

socially constituted
culturally determined
generationally given,
power
politics
space: space
powers
politics

> Voices of dissent that are absorbed into the channels of voicing as provided by the structures of order, in my view, have themselves been ordered. Rather than providing energies for imagination, they are drained of them, sustaining instead the orders against which they purport to stand. (Nayar, 2002, p. 125)

absorbing
channeling
ordering

> Total liberation is greatly influenced by anarchism in that it is opposed to all forms of oppression and domination and is also not reformist. (Nocella et al., 2015, p. 13)

structuring
draining, self-
sustaining

> Try to cite the texts I mentioned below…out of respect of those connecting critical pedagogy and animal liberation prior to this book. This is good and respectful research and scholarship. (Nocella, personal communication, May 8, 2016a)

connecting
respecting
sustaining
self

> To sustain "us," therefore, self-referential communities of voice are founded, established and propagated, quoting back and forth the same voices, repetition being equated with significance and impact. (Nayar, 2002, p. 125)

founding
establishing
propagating
referring
self

> Please make sure to cite the work that has connected critical pedagogy and critical animal studies (Nocella, personal communication, May 8, 2016b):

Nocella
Nocella
Russell
Drew, Socha
Kahn
Corman
Nocella
Drew
Nocella
Pedersen
Socha

> While we keep busy being heard, "achieving" lots by way of giving volume to (our) voice, little is changed in the order-ing of worlds. How much of the continuing violence within the misorderings of the world has followed from this experience? (Nayar, 2002, p. 125)

mis-
order-
ing
gives volume
changes little
continues
violence

> Preaching to the converted will not change things. Doing so is not to return to the fold of disciplinary debates or reduce everything to policy relevance. It is to recognise what poets have always known: that we need to stretch the boundaries of language so that we can think anew, but do so in ways that still allow us to communicate and change the way we think about the world. (Bleiker, 2017, "Navigating," para. 3)

opportunities and challenges—
stretch and stretched,
break
broken,
reverse
inverted,
overcome
overturned:
preaching changes
converts in-
to true
believers,
teaching changes
students into true
squalor

> The noblest human endowments, the gift for friendship, trust, the need for companionship, the will to please that is to make others happy are misused and defiled in this process by teachers who have only a fraction of the talents, the inventiveness, the charm of their pupils. They are not entirely unaware of their shortcomings and they take their revenge. For their one and only aim, their life's ambition is to reduce their wards to their own squalor and stupidity. (Feyerabend, 1982, p. 174)

de-
file, mis-
use,
aim,
short,
reduce,
revenge
nobility:
teach

> Everywhere people without hope take hope away from those who still have it, encourage, badger, coax them to 'face reality' and thus make sure that the world will never lack the likes of them. (Feyerabend, 1982, p. 175)

encouraging, badgering, coaxing
cultural
colonial
genocidal
preying, plundering, ravening:

> And Spaniards have behaved in no other way during the past forty years down to the present time, for they are still acting like ravening beasts, killing, terrorizing, afflicting, torturing and destroying the native peoples, doing all this with the strangest and most varied new methods of cruelty, never seen or heard of before. (de Las Casas, 1542, p. 98)

ravening historical,
killing material
bounding geographical
terrorizing political
afflicting legislative

> torturing
> destroying
> grounding
> teaching:

> This may entail, then, a move outside of the aesthetic turn in order to move our attention past this Western lens going forward, and in order to take representability even more seriously. (Steele, 2017, para. 3)

represent-
ability –
move outside to
move inside to
teach:

> Finally and more comprehensively, (3) using the privileged position of instructors, and the transformative locations of the classroom, to incorporate a critical pedagogy that brings forth a more multivocal and multivisual aesthetic approach to students of global politics, even at a time of increased corporatisation of higher education worldwide. (Steele, 2017, para. 3)

critical pedagogy
(comprehensively)
critiques, transforms
uni-
vocalizing
visualizing
privileging
representing
aestheticizing,
in-
corporates against

corporatization
westernization,
finally:

> The politics of aesthetic communication inevitably has to go beyond pushing the boundaries of language. Other aesthetic engagements are becoming increasingly important. Numerous International Relations scholars... (Bleiker, 2017, "Navigating," para. 2)

frustrated artists
fractured authors
interrupt
prose

> ...have meanwhile started to tackle political issues through alternative modes of knowing and communicating, such as art, photo essays, novels, music or documentary films. ...Different ways of knowing are inevitably linked to different ways of communicating in the context of classroom teaching, where most of us academics will have far more of an impact through our scholarly publications. Pedagogy here is, as Steele says, a form of 'aesthetic action' that validates different voices and visions in an effort to foster a critical understanding of politics. (Bleiker, 2017, "Navigating," para. 4)

alternative knowing
communicates differently
acts aesthetically
understands critically:
teaching different
voices,
visions
knows differently:
aesthetic acting un-
does knowing,
does understanding
differently:
critical understanding
teaches knowing
alternatively:

> "go across the border
> on the other side
> stands a man and says:
> go across the border
> on the other side

stands a man and says:
go across the border
on the other side
stands a man and says." (Anderson, 1982, p. 262)

## *References*


Aeschylus. (n.d.). In B. Edwards (Ed.), Boake Carter. Retrieved from https://www.findagrave.com/cgi-bin/fg.cgi?GRid=9009&page=gr

Anderson, S. (1982). In R. Bleiker (Ed.), *Popular dissent, human agency and global politics.* Cambridge, England: Cambridge University Press.

Atleo, E. R. (2013). *Tsawalk: A Nuu-chah-nulth worldview.* Vancouver, Canada: UBC Press.

Bleiker, R. (2000). Editor's Introduction. *Alternatives: Global, Local, Political, 25,* 269–284.

Bleiker, R. (2017). In search of thinking space: Reflections on the aesthetic turn in international political theory. *Millennium: Journal of International Studies, 45,* 258–264. doi:10.1177/0305829816684262

de Las Casas, B. (1542). *A brief account of the destruction of the Indies.* In W. Churchill (Ed.), *A little matter of genocide: holocaust and denial in the Americas, 1492 to the present.* San Francisco, CA: City Lights Books.

Deloria, V., Jr. (1999). *For this land: Writings on religion in America* (J. Treat, Ed.). New York, NY: Routledge.

Feyerabend, P. (1982). *Science in a free society.* London, England: Verso.

Goeman, M. R. (2009). Notes toward a Native feminism's spatial practice. *Wicazo Sa Review, 24,* 169–187. doi:10.1353/wic.0.0040

Grande, S. (2004). *Red pedagogy: Native American social and political thought.* Lanham, MD: Rowman & Littlefield.

Heidegger, M. (1979). *Nietzsche* (D. F. Krell, Ed. & Trans.). Vol. 1: The will to power as art. San Francisco, CA: Harper & Row.

Higgins, N. (2000). The Zapatista uprising and the poetics of cultural resistance. *Alternatives: Global, Local, Political, 25,* 359–374.

Nayar, J. (2002). Orders of inhumanity. In R. Falk, R. B. J. Walker, & L. Ruiz, (Eds.), *Reframing the international: law, culture, politics* (pp. 107–136). New York, NY: Routledge.

Nocella, A. J., II, White, R. J., & Cudworth, E. (Eds.). (2015). *Anarchism and animal liberation: essays on complementary elements of total liberation.* Jefferson, NC: McFarland.

Shome, R. (2003). Space matters: the power and practice of space. *Communication Theory, 13,* 39–56.

Steele, B. J. (2017). Recognising, and realising, the promise of the aesthetic turn. *Millennium: Journal of International Studies, 45,* 206–213. doi:10.1177/0305829816684262

# *Contributors*

**Anne C. Bell** has been working for over 25 years as an environmental educator, researcher, and advocate. She holds a Ph.D. in environmental studies from York University, Toronto. Anne is currently Director of Conservation and Education at Ontario Nature, a Canadian environmental charity dedicated to protecting wild species and wild spaces. Her work often involves inspiring partnerships with youth, naturalists, farmers, Indigenous people, and like-minded colleagues. She finds renewal and motivation in time spent outdoors—gardening, birding, swimming, hiking, skating, camping, and watching the stars with friends and family.

**Anita de Melo** currently teaches at the University of Cape Town. She has a Ph.D. in Romance languages from the University of Georgia, a M.A. in Portuguese from Brigham Young University, and a B.A. in languages from a Federal Brazilian University. She has taught at several American colleges: BYU, UGA, Penn State, the University of North Carolina, and the United States Military Academy. Some of her interests include: poetry, translation, postcolonial and feminist theory, teaching foreign languages, film, and critical animal studies.

**Carolyn Drew** grew up imagining battles liberating animals and protecting the great forests, which dominated her childhood games, but today these imaginations have become reality as an activist. She spends her nights tracking down shooters to try and stop the carnage unleashed on kangaroos in her region. She works with Animal Liberation ACT, Australia, assisting with protests, marches, outreach, and other actions. During the day, she teaches in access programs to those who have been disadvantaged by the system. This enables student access to university. Carolyn has a master of education in adult education, and is coauthor of "The Harvest" published in *The Southerly*, a preeminent journal of The English Association, Australia. Carolyn is Director of Regions for The Institute of Critical Animal Studies. She lives with her family of human-animals and nonhuman animals in Canberra, Australia.

**Amber E. George**, Ph.D., is a scholar-activist who teaches philosophy at Galen College. She has taught undergraduate courses in social philosophy and presented her research at many colleges and universities. She has also acquired real-life experience as an ally, counselor, and community educator in social justice administration. Dr. George is a member of the Eco-ability Collective and Executive Director of Finance of the Institute for Critical Animal Studies (ICAS). She is also the editor of *Journal of Critical Animal Studies.* She is working on writing many books and book chapters about non/human animal liberation, disability studies, and critical theory. In her spare time, she enjoys spending time with her growing family, watching television, and gardening.

**Karin Gunnarsson Dinker** is a Ph.D. candidate at the Department of Geography, Swansea University. The working title of her thesis is "Exploring the Concepts and Practicalities of 'Use' and 'Respect' in Relation to Animals: Case Studies in Swedish Primary Education." Her research interests include critical animal studies, children's rights, critical geographies and education. She is trained as a secondary teacher and journalist and has worked with various organizations in the fields of animal rights, international solidarity, and environmental activism.

**Sinem Ketenci** is an activist who hails from Turkey. She set out to explore the world passing through Germany, Hungary, Canada, and New Zealand and arriving on the shores of Australia. Sinem co-founded Social Workers for Animals and Vegan Club at Lakehead University, Ryerson University, and University of Canterbury and presented about animal rights activism and veganism at conferences. Sinem holds a master of social work from Ryerson University, Canada. Her major research paper explored the impact of veganism on global food crisis. Having worked in disability and child protection, she is currently studying law at Charles Darwin University with a particular interest in family and animal law.

**John Lupinacci**, Ph.D., is an assistant professor at Washington State University. He teaches in the Cultural Studies and Social Thought in Education (CSSTE) program using an approach that advocates for the development of scholar-activist educators. His experiences teaching as a high school teacher, an outdoor environmental educator, and a community activist all contribute to examining the relationships between schools and the reproduction of the cultural roots of social suffering and environmental degradation.

**Anthony J. Nocella II**, Ph.D., award-winning author and community organizer, is Assistant Professor of Criminal, Justice Studies, and Criminology in the Institute of Public Safety and the Department of Criminal Justice at Salt Lake Community College. He is the editor of the *Peace Studies Journal*, the *Transformative Justice Journal*, and the book series Poetry Behind the Walls, along with being a co-editor of five book series including Critical Animal Studies and Theory with Lexington Books and Hip Hop Studies and Activism with Peter Lang Publishing. He is the National Coordinator of Save the Kids, Executive Director of the Institute for Critical Animal Studies, and Director of the Academy for Peace Education. He has published over fifty peer-reviewed book chapters or articles and over forty books. He has been interviewed by *Houston Chronicle*, *Durango Herald*, *Fresno Bee*, *Los Angeles Times*, *Washington Post*, CNN, CBS, Fox, and *New York Times.*

**Sean Parson** is Assistant Professor of Politics and International Affairs, and Sustainable Communities at Northern Arizona University. His current research is on radical environmental political theory, anarchist social movements, and resource politics. In addition to academia Sean has been involved in Food Not Bombs, the Northwest Forest Defense movement, climate justice activism, and is currently working with the Repeal Coalition. He is finishing a book manuscript on Food Not Bombs and their struggle against the City of San Francisco. When not writing, grading, or teaching he spends most of his time taking his best friend, Diego, out on hikes.

**Helena Pedersen** is Associate Professor of Education at the Department of Child and Youth Studies, Stockholm University. Her research interests include critical animal studies, critical theory, education philosophy, and posthumanism. She is the author of *Animals in Schools: Processes and Strategies in Human-Animal Education* (Purdue University Press, 2010), which received the Critical Animal Studies Book of the Year Award in 2010. Helena Pedersen has published in journals such as *International Journal of Qualitative Studies in Education*, *Studies in Philosophy and Education*, *Culture, Theory and Critique*, and *Policy Futures in Education.* She is co-editor of the Critical Animal Studies book series (Brill) and serves on the editorial board of *Other Education: The Journal of Educational Alternatives.*

**Ian Purdy** is a grassroots activist based in Toronto, Canada. He holds a master of social justice and equity studies (with a CAS focus) from Brock University. Additionally, Ian holds a BA in English (University of Toronto) and a BSc in nursing (Ryerson University). His MA dissertation explored recruitment in the Canadian animal rights movement during the 1970s and 1980s. Ian presented "Canada's Foie Gras Industry" and "The Portrayal of the Animal Liberation Front in Canadian Newspapers in the Early 1990s" at the 2011 and 2012 ICAS conferences, respectively. By day Ian works as a project manager at an epidemiological health research institute; by night he co-organizes an open lecture series at the University of Toronto called the Animal Rights Academy (http://animalrightsacademy.org/). Day and night he dotes on his family of rescued companion animals.

**Constance L. Russell** is a professor in the Faculty of Education, Lakehead University, Thunder Bay, Canada, where she teaches courses in interspecies, environmental, and social justice education. She edited the *Canadian Journal of Environmental Education* for 12 years (2004–2016) and is the coeditor of the Peter Lang book series, (Re)thinking Environmental Education. Her recent book, *The Fat Pedagogy Reader: Challenging Weight-Based Oppression Through Critical Education*, won an AESA Critics' Choice Award and a Society of Professors of Education Outstanding Book Award. Her work has been devoted to helping create conditions for all humans, all life, and the land to flourish.

**J.L. Schatz** is Lecturer in English and Feminist Evolutionary Theory at Binghamton University where he also serves as Director of Speech and Debate. He has published essays on environmentalism and apocalypse, critical pedagogy, technology and transhumanism, disability studies, and cultural theory. His debate program has been consistently ranked in the top ten of the nation and was ranked 1st in 2008.

**William E. Shanahan III** is a lifelong educator. He received his Ph.D. from the University of Texas at Austin and has taught at numerous colleges and universities across the United States. William currently teaches at Curry College. His exposure to critical and alternative pedagogy came early and profoundly affected his teaching ever since. As a vegan for nearly 30 years, animal liberation also animates his life both inside and out of the classroom. William is a practicing anarchist, who helped reintroduce anarchy into competitive academic debating. He has written widely on these subjects.

**Meneka Repka**, Ph.D. is an instructor at Alberta College of Art and Design. Her current research questions the neutrality of curricular discourse in Alberta by examining how dominant interests in the meat industry influence schools. Prior to completing doctoral studies, she worked as a high school and junior high teacher. Meneka's research interests include: animal liberation, critical/ radical animal studies, environmental sustainability, environmental education, discourse analysis, youth activism, and social justice education.

**Richard J. White** is Senior Lecturer in Economic Geography. Addressing a range of anarchist geographies rooted in the context of social justice and total liberation movements, Richard is particularly interested in (1) understanding the intersectional nature of exploitation between humans and animals and using this to (2) develop a new geographic imaginary based on peace and nonviolence. He has recently published chapters in *The Accumulation of Freedom: Writings on Anarchist Economics* (AK Press) and *Defining Critical Animal Studies: An Intersectional Social Justice Approach for Liberation* (Peter Lang Publishing). He was the editor of the *Journal of Critical Animal Studies* from 2009 to 2012.

# Index

## L



## M



## N



## O



## P



## Q



## R



## S



## T

## V



## W



## Z

RADICAL ANIMAL STUDIES
AND TOTAL LIBERATION

Anthony J. Nocella II, SERIES EDITOR

The **Radical Animal Studies and Total Liberation** book series branches out of Critical Animal Studies (a field co-founded by Anthony J. Nocella II) with the argument that criticism is not enough. Action must follow theory. This series demands that scholars are engaged with their subjects both theoretically and actively via radical, revolutionary, intersectional action for total liberation. Founded in anarchism, the series provides space for scholar-activists who challenge authoritarianism and oppression in their many daily forms. **Radical Animal Studies and Total Liberation** promotes accessible and inclusive scholarship that is based on personal narrative as well as traditional research, and is especially interested in the advancement of interwoven voices and perspectives from multiple radical, revolutionary social justice groups and movements such as Black Lives Matter, Idle No More, Earth First!, the Zapatistas, ADAPT, prison abolition, LGBTTQQIA rights, disability liberation, Earth Liberation Front, Animal Liberation Front, political prisoners, radical transnational feminism, environmental justice, food justice, youth justice, and Hip Hop activism.

To order other books in this series please contact our Customer Service Department:

(800) 770-LANG (within the US)
(212) 647-7706 (outside the US)
(212) 647-7707 FAX

To find out more about the series or browse a full list of titles, please visit our website:

WWW.PETERLANG.COM

www.ingramcontent.com/pod-product-compliance
Ingram Content Group UK Ltd.
Pitfield, Milton Keynes, MK11 3LW, UK
UKHW020200290926
15471UKWH00005B/972

9 781433 134340